WATERFRONT MANHATTAN

KURT C. SCHLICHTING

WATERFRONT MANHATTAN

FROM HENRY HUDSON TO THE HIGH LINE

JOHNS HOPKINS UNIVERSITY PRESS

Baltimore

Johns Hopkins University Press
2715 North Charles Street
Baltimore, Maryland 21218-4363
www.press.jhu.edu

Library of Congress Cataloging-in-Publication Data
Names: Schlichting, Kurt C., author.
Title: Waterfront Manhattan : from Henry Hudson to the high line / Kurt C.
    Schlichting.
Description: Baltimore : Johns Hopkins University Press, [2018] | Includes
    bibliographical references and index.
Identifiers: LCCN 2017037462| ISBN 9781421425238 (hardcover : alk. paper) | ISBN
    9781421425245 (electronic) | ISBN 1421425238 (hardcover : alk. paper) | ISBN
    1421425246 (electronic)
Subjects: LCSH: Harbors—New York (State)—New York—History. | New York
    Harbor (N.Y. and N.J.)—History. | Manhattan (New York, N.Y.)—Economic
    conditions. | BISAC: HISTORY / United States / General. | SCIENCE /
    History. | ARCHITECTURE / Urban & Land Use Planning. | TRANSPORTATION /
    Ships & Shipbuilding / History.
Classification: LCC HE554.N7 S35 2018 | DDC 387.109747/1—dc23
    LC record available at https://lccn.loc.gov/2017037462

A catalog record for this book is available from the British Library.

*Special discounts are available for bulk purchases of this book. For more information, please contact
Special Sales at 410-516-6936 or specialsales@press.jhu.edu.*

To Mary with love:

with you all has been possible.

Contents

Preface   ix

1 Growth, Decline, and Rebirth   1

2 Water-Lots and the Extension
of the Manhattan Shoreline   16

3 The Ascendency of the Port of New York   38

4 New York's Waterway Empires   62

5 The Social Construction of the Waterfront   87

6 The Port Prospers, the Railroads Arrive,
and Congestion Ensues   109

7 The Public and Control of the Waterfront   131

8 Crime, Corruption, and the Death
of the Manhattan Waterfront   157

9 Rebirth of the Waterfront   188

Notes   213

Index   231

Preface

My engagement with Manhattan dates back to childhood trips from Connecticut to "the City" with my grandmother or extended family on the old New York, New Haven & Hartford Railroad. Later, in high school and college, the lure of New York's energy beckoned. When starting graduate school at New York University in 1970, I moved to the Lower East Side to share a loft on East 10th Street. The methadone clinic across the street drew junkies, who filled the stoops during the day. Street people lived in Tompkins Square, and the squats on the adjacent avenues provided a real-time view of the "urban crisis."

≈≈ June 2017

The city's vibrant Hudson River Park could not have looked more beautiful, filled with thousands of people enjoying the breathtaking views of the Hudson and, across the water, to the brand-new waterfront in Jersey City and Hoboken. Frisbees and balloons filled the air. Upscale parents pushed strollers along the shoreline. Tourists from all over the world joined New Yorkers to enjoy Manhattan's reclaimed meeting of land and water.

At Stuyvesant Cove Park, the path winds along the East River, from East 20th Street south to the Williamsburg Bridge and then down to the Battery. Across the East River, gleaming new apartment buildings line the shore in Long Island City and Williamsburg, two city neighborhoods in the midst of a renaissance. Farther to the south, Brooklyn Bridge Park offers spectacular views of Lower Manhattan, crowded with gleaming new office buildings and home to Wall Street, the center of world finance.

In Greenwich Village, luxury buildings on Charles Street cast deep shadows on Charles Lane, the once-infamous "Pig Alley" where drunks from the longshoremen's bars on West Street slept off a bender. Directly across the street in Hudson River Park, where ships from all over the world once tied up, two new rebuilt piers offer space for sunbathers to lounge and catch the gentle breezes. North at the Chelsea piers, people golf and rent sailboats.

～～ June 1971

As usual, some friends and I ended up late at night at the White Horse Tavern on Hudson Street in Greenwich Village, looking for the ghost of Dylan Thomas. I faced the dangerous prospect of walking across town to my apartment on East 10th Street. On the corner of my block, even the prostitutes who worked the fleabag hotel on 2nd Avenue would be afraid to be out. Just two weeks earlier, at about 8:30 p.m., I was in my third-floor loft and heard a loud banging in the hall. My neighbor and I looked down the stairwell and saw three guys with a 4×4 trying to bang down the door of an apartment on the next floor. We ran into my apartment, bolted the door, and called the police. They arrived in a few minutes and arrested the would-be thieves. We watched as the miscreants were loaded into a squad car, which drove away with the 4×4 sticking out of the trunk.

Back at the White Horse, John and the other Jersey boys offered me a ride across town before they hit the Holland Tunnel and home. A problem immediately arose. Earlier in the evening, John had parked on West Street, under the elevated highway, and now we faced the prospect of walking three blocks down Perry Street to West Street at 2 a.m. in the morning. We were all terrified—five 20-something rugby players fearful to walk through an apocalyptic streetscape in Greenwich Village to the crumbling understructure of the West Side Highway, the abandoned warehouses, and the dilapidated piers along the Hudson River. We survived the journey, and I have lived to marvel at the rebirth of the Manhattan waterfront and the city of New York.

This book views the history of the Manhattan waterfront as a struggle between public and private control of what is New York City's most priceless asset. From colonial times until after the Civil War, the city ceded control of the waterfront to private interests, for the latter to build a maritime infrastructure. With only finite space available, despite the made-land along the shore, a battle ensued among shipping interests, the railroads, and the ferries for access to the waterfront. The public was excluded from the waterside.

In the second half of the nineteenth century, the city of New York regained control of the waterfront, and Manhattan and the Port of New York remained preeminent until after World War II. A whirlwind of forces beyond the control of either public or private interests followed: crime on the waterfront, the revolution brought about by containerized shipping, the deindustrialization of New York City and its metropolitan region, the changing ethnic and racial makeup of the waterfront neighborhoods, and the ecological vulnerability of the extension of the Manhattan shoreline out into the Hudson and East Rivers. Hurricane Sandy provided a devastating warning of the consequences of global warming.

New York recovered from the depths of darkness in the 1970s, '80s, and '90s. The Manhattan shoreline has been reimagined as a place for the public to enjoy and the affluent to live. A complicated web of public and private initiatives has transformed the waterfront from the center of the city's maritime world to a magnificent setting for leisure and recreation.

My two previous books for Johns Hopkins University Press drew upon archival materials at the New York Public Library, including a treasure trove of historic maps of Manhattan, extending back to the original Dutch settlement. During a visit to the map room with a group of my students from Fairfield University, Matt Knutson, then curator of the Map Division and now head of humanities research, shared a series of historic maps created by New York City's Department of Docks in the 1870s. The maps detailed the water-lots granted by the city to private owners, who extended Manhattan Island out into the surrounding rivers and then built the piers that were essential to the rise of the Port of New York. Over time, the water-lot grants led to the addition of thousands of acres of made-land, a pivotal part of Manhattan's history. As an idea for a book materialized, I realized that more research was needed, which I pursued in the magnificent resources of the New York Public Library.

At a meeting in Baltimore with my editor at Johns Hopkins University Press, Robert Brugger, I discussed my preliminary research and presented a broad outline for a study of the Manhattan waterfront, from its Dutch settlement to the present day. As always, he encouraged me to continue the research and, when ready, to prepare a précis for his review. This book grew out of that meeting, and Bob's steadfast support has been crucial.

Jameson Doig, emeritus professor at Princeton University and author of *Empire on the Hudson*, read the final draft of the book. His insightful comments strengthened the manuscript. Jim's seminal work on the Port Authority sets the standard for any study of New York Harbor and the metropolitan region, from the twentieth century to the present day. An interview with Dan Pastore of the Port Authority of New York and New Jersey provided valuable insight into the world of the container ship. Once again, it has been a pleasure to work with the consummate professionals at Johns Hopkins University Press: Elizabeth Demers, senior editor; Juliana McCarthy, managing editor; and Lauren Straley, editorial assistant. Kathleen Capels's copyediting has been meticulous. I appreciate her insight and careful questions necessary to complete the book.

I hold the E. Gerald Corrigan '63 Chair in Humanities and Social Sciences at Fairfield University. Dr. Corrigan is a past president of the Federal Reserve Bank of New York. The Corrigan Chair provided crucial research funding for this book, including numerous trips to archives in New York, Philadelphia, and the Library of Congress in Washington, DC.

At each of these archives, I relied on the assistance of truly dedicated professionals, whose knowledge of the materials proved invaluable. Matt Knutzen, at the New York Public Library, led the creation of the brilliant Map Warper website that provides public access to thousands of digitized historical maps of Manhattan. The maps are georeferenced, and I included many as base layers in the GIS maps created for this volume. Thomas Lannon, curator of the Manuscripts and Archives Division at the New York Public Library, was once again of great assistance, as he was with my William Wilgus book. Kenneth Cobb, at the New York City Municipal Archives, assisted with the records of the water-lots, compiled by the New York Department of Docks in the 1870s. The staff at the New York Historical Society also helped by allowing me to read some of the original water-lot grants, handwritten on parchment. Records of some of the earliest Quaker ship owners in New York are in the collections of the G. W. Blunt White Library at Mystic Seaport in Connecticut, and the staff there were very accommodating during numerous visits.

Michael Kuhn, a Fairfield University undergraduate and then a graduate research assistant, diligently undertook some of the most tedious work needed to assemble the research material into a coherent form. Michael accompanied me to the New York Public Library numerous times, as well as walked the streets of Greenwich Village, historic maps in hand, to study the remaining nineteenth-century streetscapes adjacent to the Hudson River. Steve Evans at the Fairfield University Media Center edited many of the digital maps from the New York Public Library and made the past visible.

This and my previous books for Johns Hopkins University Press would never have been possible without the persistent and gentle support of my wife Mary, aka Dr. Mary Murphy of the English Department at Fairfield University. Without complaint, she patiently read every draft of every line. Her comments were always, always perceptive, and I followed her wise counsel!

For additional resources and maps relating to *Waterfront Manhattan*, see https://digitalhumanities.fairfield.edu/Waterfront_Manhattan.

WATERFRONT MANHATTAN

[1]

Growth, Decline, and Rebirth

Manhattan Island, just 22.4 square miles in total, 23 miles long, and, at 14th Street, 2.3 miles wide, would emerge as the most populous place in the United States and the country's preeminent city. Until consolidation took place in 1898, the island alone constituted the city of New York. Brooklyn, across the East River on Long Island, was a separate city, and the outer boroughs remained largely rural farmland.

At the time of the first US census in 1790, the Atlantic seaboard cities—Boston, Newport, Philadelphia, and Baltimore—vied to capture the new country's maritime commerce. By 1850, New York City's population had soared to over half a million, and it reigned as the leading metropolis in the country. The growth and prosperity of the port explains the dominance of the city in the nineteenth century.

Manhattan became a global city by building maritime links across the oceans, along the Atlantic coast, and inland to the Midwest and New England. The shorefront of Manhattan Island, the boundary between water and land, served for hundreds of years as the center for the trade, shipping, and commerce of the nation as the port became the busiest in the world. No other activity took precedence over commerce. The shore remained a place apart, a world of piers and docks; longshoremen; and ships, tugboats, and ferries filled with cargo and freight; as well as a site where millions of immigrants entered the Promised Land.

Nature provided New York with a sheltered harbor, but to flourish, the port required a complex maritime infrastructure of wharves, piers, warehouses, and waterfront streets to facilitate the transfer of goods and people between water and land. A busy port depended on a large workforce of itinerant day laborers who lived nearby and would be available on the docks, in all weather, as ships arrived and departed.

1

A prosperous seaport needed substantial investment in its waterfront infrastructure to provide ships and merchants with access at reasonable costs. New York faced a challenge here: to find the necessary capital to build and expand its maritime infrastructure. In the seventeenth and eighteenth centuries, the city's government did not have the responsibility or the fiscal resources to develop needed port facilities. To build this infrastructure, the city gave away a precious asset: the land under the water around the lower part of Manhattan Island. The municipal government awarded these water-lots to private individuals to build wharves and piers, surrendering public control of the waterfront.

For over 250 years, private enterprise ran the waterfront; the city played merely a peripheral role. By the end of the Civil War, chaos threatened the port's dominance. In 1870, the city and the state created the Department of Docks to exercise public control and rebuild the port's maritime infrastructure for the new era of steamships and ocean liners. A hundred years later, technological change—in the form of container shipping and jet airplanes—rendered Manhattan's waterfront obsolete within an incredibly short time span. The maritime use of the shoreline collapsed, mirroring the near death of the city of New York in the 1970s. Ships disappeared and abandoned piers and empty warehouses lined the waterfront.

The city slowly and painfully recovered. The vacated waterfront allowed visionaries and planners to completely reimagine a shore lined with parkland. Along the revitalized waterfront, luxury housing has transformed the neighborhoods where Irish longshoremen once lived. A few remaining piers offer spectacular views of the city's waterways. The rebirth has been driven by complex private/public partnerships, with the city of New York playing only a peripheral role. The contentious question of private versus public control of the waterfront remains a continuing issue in the twenty-first century.

New York Harbor

To the south of Manhattan Island lies New York Harbor, one of the greatest natural harbors in the world, providing deep but sheltered anchorage from all but the most violent ocean storms (map 1.1). Both Staten Island and the Brooklyn section of Long Island form a natural barrier between the harbor and the Atlantic Ocean. The waters of the

Upper Bay offer space for hundreds of ships to anchor. Between Staten Island and Brooklyn, the Narrows forms a deepwater entrance. From the Upper Bay, the Hudson River leads north to Albany and to waterways across the state that reach the Great Lakes and the Midwest. The East River, a tidal estuary, connects to Long Island Sound and provides a protected passage to New England.

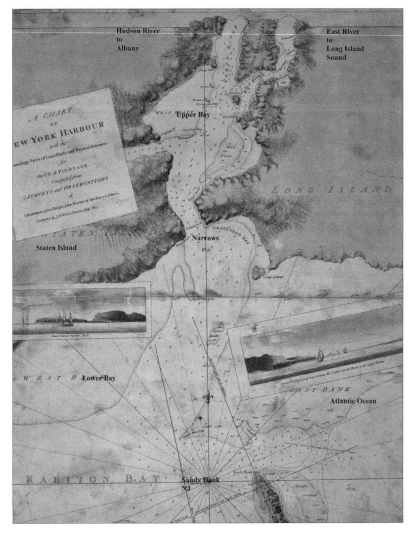

Map 1.1. New York Harbor, in 1779, showing the Narrows, the Upper Bay, and lower Manhattan Island. *Source*: "A Chart of New York Harbor, with the Soundings . . . ," 1779, Map Division, New York Public Library

Land sloping gently down to the sea, deep water extending almost to the shoreline, and a minimal rise and fall of the tide offered an ideal location for maritime trade. Both the Hudson and East Rivers are tidal estuaries, and twice a day the tide floods (up the harbor and the rivers) and ebbs (down the harbor to the ocean). Sailing ships took advantage of the tidal currents to assist them in entering or departing from the harbor.

When the first Dutch ships arrived in New Amsterdam in 1624, they anchored in the East River, people and cargo were rowed ashore in small dories. The reverse took place as a ship readied for departure. Once the volume of shipping increased, the use of dories proved to be time consuming and expensive. A more efficient system involved building wood or stone structures: wharves (paralleling the shorefront), or piers (extending from the shoreline out into the water). Both simplified the unloading and loading of ships moored to the wharves or piers. On the shore adjacent to these structures, waterfront streets facilitated the transportation of goods and people inland.

〰 Building Manhattan's Waterfront Infrastructure

The English captured New Amsterdam from the Dutch in 1664 and renamed the city New York. Two charters established the city of New York as a corporation—with its own government, consisting of a Common Council and a mayor—distinct from the colony of New York. The 1686 Dongan Charter and, later, the Montgomerie Charter of 1730 legally defined the waters around Manhattan Island to be the property of the city. The Montgomerie Charter specified that the city owned all of underwater land extending out from the island for 400 feet beyond the low-water line. It created a resource of immense value—water-lots adjacent to the shorefront—and gave the Corporation of the City of New York the legal right to grant the water-lots to private citizens.

The city used the water-lots to attract private waterfront investment. First the English governors, and later the city's Common Council, granted water-lots to individuals, often the politically connected well-to-do who owned shoreline property. In return for a water-lot grant, the private owners had a legal obligation to build a bulkhead out from the existing shoreline into the river, which would serve as a wharf, as well as to construct a new city waterfront street for public use. They could also build a pier out into the surrounding rivers. Such requirements under the grants, however, did not include specific construction details. Fees paid to use the wharves and piers went to these private owners, not the city. In addition, the grantees filled in their water-lots, creating made-land along the shore that they could then develop. On this new made-land, the owners constructed warehouses, counting-houses, tenements, and factories. The use of water-lot grants and private investment ceded control of the waterfront from the city to private interests. The city could revoke the grants, but seldom did, and it exercised only weak regulatory power over the city's most valuable asset: direct access to the greatest natural harbor in the world.

Thus the shoreline of the city moved out into the Hudson and East Rivers. Map 1.2 illustrates the water-lots and made-land in Lower Manhattan: along the Hudson River from Greenwich Street, the original shoreline, to the new wharves on West Street. This waterfront real estate, created on made-land, belonged to the grantees, not to the city of New York, and today it is among some of the most valuable real estate in the world. For 300 years, the granting of water-lots enlarged the island of Manhattan by 3.57 square miles, adding 2,286 acres (map 1.3).[1]

As the port grew, the city continued to use water-lot grants to facilitate the building of additional infrastructure along the East and Hudson Rivers, but the wharves alone could never meet the demand as the volume of shipping increased. By the Civil War period, captains, ship-owners, and city merchants complained vociferously about the lack of pier space, the rotting wharves, poor maintenance, overcrowded waterfront streets, inadequate storage, corruption, and the gangs of thieves who pillaged the waterfront. Thunderous newspaper editorials warned that without dramatic improvements, New York City would lose maritime business to Boston, Baltimore, or Philadelphia. Private ownership had failed to build, maintain, and upgrade the essential infrastructure and threatened the port's maritime supremacy.

In 1870, the city and the state of New York passed legislation to establish a public agency, the Department of Docks, to reassert public control over the waterfront. At great expense, the city purchased the wharves and piers from the private owners and set out to rebuild the waterfront, operating the port for the public good, not for private gain.

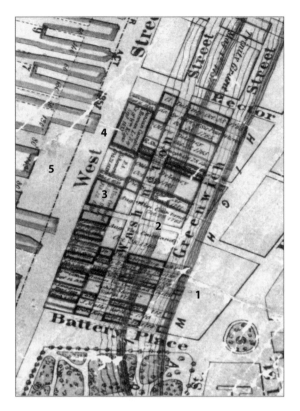

Map 1.2. Water-lots on the Hudson River, from the Battery to Rector Street. The numbers on the map indicate the original Manhattan shoreline, with high water at Greenwich Street (1); the first water-lot grants, in the 1760s, from Greenwich Street to Washington Street (2); the second water-lot grants, in the 1830s, from Washington Street to West Street (3); West Street, a new waterfront street (4); and wharves and piers along the Hudson River (5). *Source*: Created by Kurt Schlichting. Source GIS layer: water lots, 1870 Department of Docks map, New York Public Library, Map Warper.

The Department of Docks managed to accommodate vast changes in the harbor as the size of both cargo and passenger ships increased dramatically. The tables turned again at the beginning of the twentieth century, when shipping companies, business leaders, and even the corrupt longshoremen's union warned of the coming demise of the harbor because of its mismanagement by the department. A whirlwind of change did arrive, but maladministration by New York City's Department of Docks played only a tangential role.

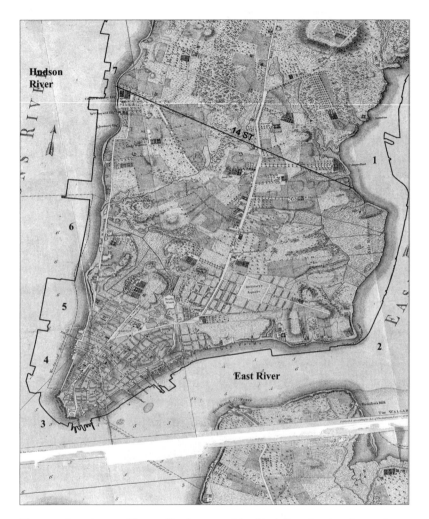

Map 1.3. A 1776 map of Lower Manhattan, with a black line showing the modern shoreline. The numbers on the map indicate Stuyvesant Cove/Stuyvesant Town (1); Corlears Hook, on the Lower East Side (2); Battery Park (3); Battery Park City (4); the Freedom Tower/World Trade Center (5); Pier 40, on Canal Street (6); and Chelsea (7). *Source*: Created by Kurt Schlichting. Source GIS layers: 1776 Ratzer map, New York Public Library, Map Warper; shoreline map, New York City Planning Department.

≈≈ Waterway Empires

Taking advantage of nature's gifts, the Port of New York came to dominate three commercial waterway territories: the first, routes across the North Atlantic Ocean to and from Europe; the second, a coastal shipping empire to the American South and the Caribbean; and the third, inland transportation up the Hudson River to upstate New York, the Great Lakes, and Chicago, as well as up the East River, through Long Island Sound, to Connecticut, Rhode Island, and New England. The city's most important international links stretched across the North Atlantic from New York to Liverpool in England and Le Havre in France. In the early nineteenth century, with relentless determination, the city's shipowners captured the North Atlantic waterway. New York Harbor served as the center of the country's export and import trade. All of New York's maritime rivals fell far behind.

On October 27, 1817, a woodcut ad appeared in a New York newspaper, announcing that as of the first week in January 1818, a new packet line of ships would begin monthly sailings from New York and, simultaneously, from Liverpool. The Black Ball Line, founded by five of New York's leading Quaker shipowners and merchants, guaranteed service from an East River pier regardless of the weather, including if there was a full cargo or not. Not only did the owners of the Black Ball Line, at great financial risk, begin a transportation revolution, but the packets also created a new communications link across the North Atlantic. In addition to cargo and first-class passengers, the packet ships carried commercial mail. An international trading system required the fastest communication possible between New York and Europe. Letters of credit, private banking correspondence, bills of sale, and specie had to be sent back and forth, and the packet lines provided fast, scheduled service.

Rival packet lines to Liverpool followed: the Red Star Line and the Blue Swallowtail Line in 1822 and, later, the Dramatic Line (1835) and the New Line (1839). Packet ships left New York and Liverpool every week of the month, every month of the year. No other port in the United States offered anywhere near this level of service. The Port of New York had packet lines not only to Liverpool, but also two lines to London and three to Le Havre, strengthening the city's hold on the ocean pathway to Europe (see table 4.1).

New York's link to Liverpool allowed the city to control the cotton textile trade between the slave plantations in the American South and the textile mills in Lincolnshire in northwestern England, to which Liverpool provided access. The industrial revolution in England created an insatiable demand for the high-quality cotton grown in the South. For more than a century, cotton remained the most valuable product produced by the American economy, creating a direct link between slaves on the southern plantations and the nascent industrial workforce laboring in the satanic conditions of the textile mills in England. New York Harbor and its shipping industry played a pivotal role. Each year, tens of thousands of cotton bales carried north on coastal ships arrived on the piers along the East and Hudson Rivers. Once unloaded, the bales were carted along the waterfront to waiting packet ships bound for Liverpool. Textiles returned from Liverpool to be distributed throughout the country, including down the Atlantic coast to the South. Even cotton transported directly from the southern cotton ports to Liverpool arrived on ships owned or financed by New York's shipping businesses.

The city's merchant bankers developed a sophisticated system of international credit that enabled plantation owners in the South to purchase slaves and grow cotton and then consign the cotton to traders, who shipped it to England. In Liverpool, import merchants bought the cotton and then quickly sold it to the textile mills. Lastly, the swift packet ships returned from England carrying English gold and silver back to New York, for the bankers to use both to settle accounts, including the funds advanced each year to the plantation owners in the South, and, of course, to take their own substantial profits. Capital accumulated in the cotton trade enabled the New York bankers to finance the shipping industry, the city's maritime infrastructure, and the burgeoning industrial revolution in the city and the United States.

A second crucial waterway linked New York with coastal ports in the South—Charleston, Savannah, Mobile, and New Orleans—that handled the eighteenth century's key agricultural product: cotton. By 1832, New York's shipowners established packet lines to all of the major southern ports. With scheduled service, these southern packets offered fast shipment of cotton to Liverpool, with the first leg running up the coast to the Manhattan waterfront, and the second leg traveling across the North Atlantic to England.

New Yorkers owned many of the cotton ships that sailed directly from the southern ports to Liverpool. The city's merchants and bankers also provided crucial ancillary services: insurance, credit, export and import services with their Liverpool counterparts, and the settlement of international payments. The pivotal role played by New York—always at a substantial profit—infuriated the plantation owners and the southern cotton aristocracy. The maritime historian Robert Albion referred to the city's dominance as "enslaving the cotton ports."[2]

A third waterway empire ran to the north: up the Hudson River to Albany and then west by the Mohawk River to central New York. From northern Alabama to Maine's border with Canada, the Appalachian Mountains separate the Atlantic seaboard from the Midwest and, in the eighteenth and early nineteenth centuries, posed a serious challenge to travel. In upstate New York, a gap exists in the Appalachian Mountains along the Mohawk River and then across rolling countryside to Lake Erie, at Buffalo. What become known as the water-level route served as the ideal location for the Erie Canal and, later, the route of the New York Central Railroad across New York State and further west.

The Erie Canal had an enormous impact on the history of the Port of New York, linking Manhattan to Chicago, the great city of the Midwest, characterized by historian William Cronin as "nature's metropolis."[3] New York captured the bounty of the Midwest. In return, imports from Europe bound for Chicago and the central states came through the Manhattan waterfront, transferred from ocean-going packet ships to Erie Canal barges.

Another key part of the inland waterway empire, the East River, connects New York's harbor to Long Island Sound. For over 90 miles, the Sound provides a protected water route to Connecticut, Rhode Island, and Massachusetts. From the Race, which marks the eastern end of Long Island Sound, the ocean passage to Narragansett Bay is less than 30 miles. On the north side of the Sound, the Connecticut River is navigable to a point north of Hartford, and then from the Massachusetts border to the middle of Vermont. Steamboats connected the Port of New York with cities along the Connecticut coast, Narragansett Bay, and Massachusetts, extending the commercial reach of New York City.

Local needs compounded the crush along New York City's shoreline. The rivers, bays, and Long Island Sound connected the growing city with a hinterland that supplied basic needs: food, wood for cooking and heating, and building materials. All goods had to be transported by water to the Manhattan waterfront. In addition, a fleet of ferries carried thousands of passengers across the waters each day to Brooklyn, Queens, Staten Island, and New Jersey. These ferries, along with small sailing vessels, steamboats, and canal barges, competed for the limited space on the piers surrounding Lower Manhattan, adding to the congestion of people and goods on the waterfront.

∼∼ Manhattan's Waterfront Neighborhoods

For the Port of New York to grow and prosper, it needed more than nature's gifts; the port's infrastructure; and aggressive shipowners, merchants, and bankers. The thousands of ships in the harbor each year and the millions of tons of cargo that had to be moved across the waterfront required a large labor force. These men worked as day laborers, not as long-term employees, and they could never be sure of steady work. From the time of the Dutch, as each ship arrived or departed—from the largest ocean-going sailing ships to the small single-masted sloops or humble canal barges—the cry "along shore men" rang out, and hundreds hurried to the wharves and piers to secure a day's work on the docks. Loading, unloading, and moving cargo remained a labor-intensive process for centuries.

The longshoremen worked long hours, in all types of weather. A brutal system of day labor became enshrined by the shape-up, where hundreds of men stood at the pier entrance, hoping to be hired for the day. Pier bosses demanded kickbacks, loan sharks worked the docks, and prostitution and crime flourished nearby. A code of silence prevailed. Any longshoreman brave enough to speak out was often found dead, floating next to the pier with his throat slit. In the early twentieth century, corruption prevailed in the longshoremen's union, which failed to protect its members. Organized crime eventually controlled the docks and shook down the shipping companies, which paid bribes in exchange for labor peace. Cargo theft reached epidemic levels. Waterfront corruption extended to the highest level of politics in New York City and the state.

Longshoremen and other dockworkers lived a few blocks away, in the immigrant neighborhoods along the East and Hudson Rivers. Generations of immigrants, many from Ireland, worked the docks and endured lives of hardship and toil. A tribal way of life arose, where sons followed fathers down to the docks each day. On the nearby streets, the Irish waterfront laborers, their wives, and their children lived in crowded tenements. While their life was hard, many found refuge and spiritual support in the Catholic churches that served the waterfront.

After World War II, crusading journalist Malcolm Johnson and a Jesuit priest, Fr. John (Pete) Corridan, led the fight for reform on the waterfront. They battled the corrupt longshoremen's union, the International Longshoremen's Association (ILA), to rid the piers of mobsters who controlled the daily work on the docks. Johnson wrote articles in the *New York Sun* detailing the crime and corruption there, a series for which he won the Pulitzer Prize in 1949. His main source was Fr. Corridan, a priest who dedicated his ministry to helping the Irish on the Manhattan waterfront. Corridan eventually ran afoul of the Archdiocese of New York, which received generous support from Catholic executives in the shipping companies and the senior leadership of the ILA.

Corridan provided Johnson and Budd Schulberg, a Hollywood screenwriter, with the material the latter used to write the script for the 1954 Academy Award–winning movie *On the Waterfront*, which starred Marlon Brando and was directed by Elia Kazan. This movie portrayed the dark world of the waterfront and generated enormous publicity, leading to calls for reform. Yet as long as ships continued to fill the harbor, business, political, and religious leaders turned a blind eye.

Organized crime endures on the waterfront, not on the piers in Manhattan or Brooklyn, but in the giant container port in Newark, New Jersey. Sixty-plus years after the movie was first released, a nephew of a mobster earned $400,000 a year at Port Newark; he never was off the clock, even when at home in bed. Three successive Newark longshoremen's union presidents were convicted of extortion, and investigators dug up $51,900 in a longshoreman's backyard—a payoff destined for the mob.[4]

∿ A Technological Revolution and Implosion on the Waterfront

Dramatic change did come to the New York waterfront, but not through a reform of the longshoremen's union or the expulsion of gangsters from the docks. Instead, a technological revolution—in the form of a simple metal box 20 feet long, 8 feet wide, and 10 feet tall—rendered Manhattan's complex waterfront infrastructure and its system of day labor obsolete in a short period of time. The container revolution began in April 1956, when Malcom McLean, a trucking and shipping executive, loaded twenty-four containers aboard his ship, the *Ideal-X*, in Port Newark, across the Upper Bay from the Manhattan waterfront, and set sail for Houston, Texas.

In the space of twenty years, such containers completely transformed the maritime world that had dominated the economy of Manhattan Island and the waterfront for hundreds of years. This type of shipping needed port facilities with both vast spaces around the piers to store the containers and easy access to the country's new interstate highway system, neither of which existed on the crowded Manhattan waterfront. In a heartbeat, the maritime world that had occupied this area for over 350 years relocated across the Hudson River to Newark Bay in New Jersey.

On the piers, silence fell. Only a few ships arrived along the Hudson River or Brooklyn waterfront from across the Atlantic Ocean, the Midwest, New England, the American South, or anywhere else. New York's waterway empire vanished. The Department of Marine and Aviation, the successor to the Department of Docks, could not find shipping companies willing to lease piers at any cost. Longshoremen and their union struggled to preserve waterfront jobs, but nothing could hold back the wave of technological change. A way of life for the longshoremen disappeared, and their tight-knit waterfront neighborhoods changed dramatically. Along the Hudson and East Rivers, pier after pier stood abandoned and, in some cases, fell into the river. The luxury ocean-liner business faded away, as people could now travel by air to Europe or anywhere in the world. Neither public nor private enterprise, and no amount of public or private investment, could save the maritime world along Manhattan's waterfront.

In the 1960s, the Manhattan waterfront and the entire city of New York entered an extended period of decline and despair. The elevated West Side Highway, adjacent to the Hudson River piers, collapsed from a lack of funds for proper maintenance, a fitting symbol of decline. In the late 1970s, the city of New York careened toward bankruptcy. Crime, drugs, and abandonment prevailed on the waterfront and in the surrounding neighborhoods.

≈≈ Rebirth of the Waterfront in an Era of Global Warming

Slowly, and with great effort, the city recovered from decades of decline, and this rebirth included reimagining the use of Manhattan's waterfront. Its commercial use would never return but the shoreline could become parkland and recreation space, a place to be reminded daily that Manhattan is an island surrounded by waterways. Now the waterfront's most valuable asset proved to be the views from ashore, not the water-lots below the rivers.

A massive plan to rebuild the West Side Highway led to a bitter fight over the future of the waterfront on the West Side. The Westway conflict stimulated alternative plans that imagined the shoreline as a public asset. This reinvigorated shoreline would become a place for New Yorkers and tourists alike to enjoy magnificent water views and watch the sun set over the Hudson River and Upper Bay. For the wealthy, the waterfront would become a desirable place to live, and for the fashionable, a trendy locale in which to dine or visit a gallery. Despite the defeat of the Westway proposal, the shorefront still remains a contested space. Will people from all walks of life and levels of income be welcome on the revitalized private/public waterfront, or will it become a place set apart, for only the wealthy to enjoy?

The rebirth of the Manhattan waterfront has justly received universal praise. Yet in an era of climate change, three centuries of creating made-land around the island has left the city facing another daunting challenge. The filling in of Stuyvesant Cove between 14th and 26th Streets on the East River created the land where Stuyvesant Town, a huge housing complex, sits just a few feet above high water. Battery Park City and the Freedom Tower, which replaced the World Trade Center, have similar elevations above the Hudson River. Lower Manhattan's shoreline today sits offshore, some distance from the 1776 waterfront. Extensive valuable real estate stands on this made-land.

In 2012, Hurricane Sandy devastated the Manhattan waterfront and the shoreline communities throughout the city, Long Island, New Jersey, and Connecticut. The extraordinary high tide, driven by the onshore winds, caused enormous damage and flooded Battery Park City and Lower Manhattan, including the subways and the Brooklyn-Battery Tunnel. At 14th Street, a Con Edison power plant exploded, plunging

Lower Manhattan into darkness. Climate scientists predict that with rising sea levels, an ordinary hurricane or even a storm with high winds will pose the same danger as Superstorm Sandy did, putting the revitalized shoreline and the city at grave risk. Despite the success of Manhattan's waterfront rebirth, the threat posed by global warming will remain a major challenge for New York City in the twenty-first century.

[2]

Water-Lots and the Extension
of the Manhattan Shoreline

For 350 years, New York Harbor and Manhattan Island prospered in a maritime world. Across the shorefront of the island—the boundary between water and land—the trade, shipping, and commerce of the country passed, ensuring the city's prosperity (fig. 2.1).

From the beginning, control of the harbor and Manhattan's waterfront—the region's magnificent geographical resource—has always been contentious. Colonial governments had limited fiscal resources, and the need to build wharves and piers provided a serious challenge. If the city of New York were to compete with other colonial seaports, did its government have the primary responsibility to construct and maintain the essential maritime waterfront? If not, and private interests and private capital instead built the wharves and piers, what role would the city's government play? Were private interests and the public good the same on the waterfront? Would limited government oversight ensure the proper development of the port by private interests?

～～ The Atlantic World

In the seventeenth and eighteenth centuries, the European settlement of North America created an Atlantic World.[1] Three thousand miles of ocean separated Europe from the Atlantic seaboard. The voyage across the North Atlantic from Europe to America, traveling east to west, is one of the most difficult ocean passages in the world. Prevailing winds in the North Atlantic blow from the west, and in the seventeenth, eighteenth, and nineteenth centuries, sailing ships struggled against them,

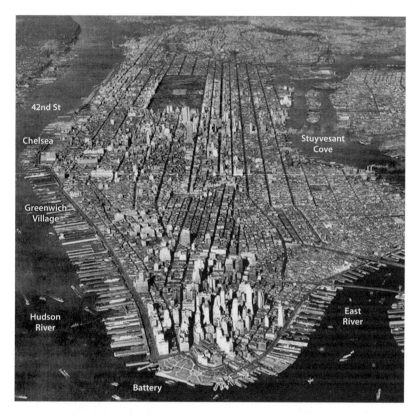

Figure 2.1. Manhattan's waterfront infrastructure, circa the 1930s. Piers line both the Hudson River shoreline (extending from the Battery to Riverside Park and West 125th Street in Harlem) and the East River shoreline (extending from the Battery to Corlears Hook, and Grand Street to East 14th Street, Stuyvesant Cove, and East 23rd Street). *Source*: New York City Municipal Archives

especially in winter, when the westerly gales blow. In US Navy Lieut. Matthew Fontaine Maury's *Explorations and Sailing Directions*, published in 1851, his wind and current data for the North Atlantic warned mariners sailing west from Europe to America to expect adverse winds in January, February, and March more than 50 percent of the time.[2]

The *Mayflower* left England on September 6, 1620, and arrived off the coast of Cape Cod in Massachusetts on November 9, a voyage of sixty-four days. Two hundred years later, crossings by the crack Black Ball Line packet ships, sailing from Liverpool to New York (1818–1853), averaged over thirty days in the winter, and some of their voyages lasted for sixty or even seventy days.[3] Sailing ships carrying immigrants took seven or eight weeks to reach New York. Sometimes they ran low on water and food, and many died en route. In our modern world, when the Atlantic Ocean can be crossed in hours by air, it is nearly impossible to imagine weeks aboard a sailing ship in winter, with snow and freezing conditions in the middle of the North Atlantic.

While the ocean voyages were extremely dangerous at times, the goal was not simply to cross the Atlantic Ocean, but to also find a way to come ashore. To reach solid ground, explorers and settlers needed to find a safe harbor. All of the early colonial cities—Boston, Newport, New York, Philadelphia, Baltimore, and Charleston—grew around inlets, rivers, and bays that provided sheltered access to the land. Besides a protected harbor, a deep anchorage, close to a gently sloping shoreline, facilitated the transfer of people and cargo from ship to shore.

Each of New York's rival colonial seaports offered some of the same geographical characteristics, so Manhattan's rise to a position of commercial ascendancy required more than just the advantages nature provided. Robert Albion, in his groundbreaking *The Rise of New York Port*, points to those advantages but also adds a crucial ingredient—the drive and ambition of the merchant class who led New York to a position of dominance: "Nature can do a good deal, but not everything, to determine the success of a seaport. Human initiative also accounts for something. . . . Geographical considerations were all important in the story of New York port. . . . At no other spot on the North Atlantic coast was there such a splendid harbor. . . . Yet even New York trailed behind less favored rivals until *local initiative* [author's emphasis] finally took full advantage of nature's bounty."[4] Albion includes a chapter on the "Merchant Princes" that provided New York with the necessary local initiative. This initiative, however, involved more than just the city. New York State, led by its hard-driving governor, DeWitt Clinton, risked millions of dollars in state funds to construct the longest canal in the world. The Erie Canal created an inland waterway between the Port of New York and a vast maritime hinterland, stretching to the Midwest.[5]

One hundred miles to the east, Newport, Rhode Island, at the mouth of Narragansett Bay, provides a deep, secure anchorage just a few miles from the ocean. Moreover, Newport harbor is closer to the Atlantic Ocean than Manhattan Island. A British piloting guide, published in

1734, included a detailed map of the New England coast from Rhode Island to New York and warned captains not to sail above 40° north latitude. Crossing the Atlantic, a sailing ship arriving at 40° north latitude could find a protected harbor in either New York or in Newport (see map 4.1). For a brief period of time, Newport rivaled Boston and New York as a center of colonial maritime commerce. The British occupied both Newport and New York during the American Revolution, along with Boston, Baltimore, and Philadelphia. Their occupation of New York lasted seven long years, from 1776 until 1783, and during that period, early in 1776, the Great Fire destroyed an estimated 25 percent of the buildings in the city. New York recovered quickly, however, and soon become the commercial center of the new country. Newport never recovered its colonial status as a preeminent port, although the other rival seaports—Boston, Baltimore, and Philadelphia—did not meekly surrender to New York's maritime ascendency. The merchants and shipowners in these cities were as ambitious as those in New York.

∼∼ New York Harbor

The Dutch chose Manhattan Island because of its superb geography, combined with access, via the Hudson River, to Albany, 140 miles to the north, which was the center of the North American fur trade. They established an all-important trading post in Albany, but they decided to build New Amsterdam on the lower tip of Manhattan Island, where the Upper Bay, just to the south of the island, forms one of the great natural harbors in the world. Once through the Narrows, which separates Brooklyn from Staten Island, ships are protected and find safe anchorage (see map 1.1). Albion describes its advantages: "New York possessed a landlocked harbor, which offered more perfect natural shelter than that of any other major American port as close to the sea. A century ago, a prominent engineer remarked that even if the sandy mass of Long Island might have no other use, it justified itself as a natural breakwater for New York. Staten Island served similar purposes. . . . Shipping in the Lower Bay or the Narrows might catch the force of a gale, but the waters around Manhattan were spared the surges of stormy seas."[6] The Battery, at the tip of Manhattan Island, is a short 16 nautical miles from the ocean. By comparison, Philadelphia is 88 miles up Delaware Bay, and Baltimore is 150 nautical miles, via Chesapeake Bay, from the Atlantic Ocean.

The Upper Bay forms a wide estuary leading to Manhattan Island, which is surrounded by the Hudson River to the west and the East River to the east. A 1779 British naval chart of the Upper Bay shows the depth of water at low tide along the Hudson and East Rivers, just off-shore around Lower Manhattan (map 2.1). Close to the shoreline, the East River's depth reaches 4 fathoms, or 24 feet, at low water. Another decided advantage to the harbor, which is actually a tidal estuary, is the minimal rise and fall of the tides, which average 4.5 feet. The deck of a ship tied to a pier along these rivers would be just 4 feet lower at the lowest point of the tide. Liverpool, an important British port located on the Mersey River, which flows into the Irish Sea and then the Atlantic Ocean, faces entirely different—and difficult—tidal conditions. Tides there rise and fall an average of 25 feet. In Liverpool, piers could not be built out into the Mersey River, as they could along the East River. At great expense, the city of Liverpool built gigantic stone basins with tidal gates. Nonetheless, ships could only enter or depart from the basins twice a day, during the few hours of high tide.

Another advantage to New York Harbor is that the currents in the Hudson and East Rivers and in the Upper Bay are not strong—only 3

Map 2.1. Lower Manhattan Island and the East River, 1779. The numbers along the shoreline represent the depth of the rivers at low tide, in fathoms. Each fathom equals 6 feet. The shoreline of Manhattan Island is ideal for a port, with the water depth at low tide being 4 fathoms (24 feet). *Source:* "A Chart of New York Harbor, with the Soundings . . . ," 1779, Map Division, New York Public Library

or 4 miles an hour on the ebb tide, flowing south toward the Narrows and the Atlantic Ocean. The one exception was Hell Gate, the narrow passage up the East River, where the river turns to the east between Randall's Island and the shoreline of New York's borough of Queens, and then connects to Long Island Sound. Here the tide is particularly strong. Submerged rocks and ledges made the passage dangerous, especially for ships sailing in light winds, with the current running. Not until the 1880s did the US government, at great expense, blast the underwater rocks and clear Hell Gate.

Given the deep water just offshore, the Dutch first used the East River as the city's port. From the Battery to Corlears Hook at Grand Street, where the river turns north to the Hell Gate, hundreds of ships crowded the wharves. Counting-houses, where the city's merchant princes worked, lined the adjacent streets. A short walk away, grog shops and cheap boarding houses provided lodging for sailors, and prostitutes worked the streets.[7]

When the British finally gained control of New Amsterdam in 1674, they recognized the advantage of using the East River waterfront as the gateway to Manhattan Island. The population of the colonial city at the tip Manhattan Island grew throughout the eighteenth century. The city of New York's population increased from 7,248 in 1723, to 13,294 in 1749, and to 21,863 in 1771, just before the American Revolution.[8]

New York did not dominate trade in the colonial period. In 1774, just before the Revolution, exports from the colonies to England totaled £804,286, with New York's share at £80,008, about 10 percent of all exports. Exports from Baltimore and Charleston, which primarily consisted of tobacco, had a much higher value than the furs, timber, and agricultural products shipped from the city of New York. New York's imports were significantly higher, totaling £437,937, or 28.6 percent of all imports.[9]

≈≈ Control and Development of the Waterfront

With the English in control, the city's colonial government first sorted out legal issues surrounding both the ownership of property on Manhattan Island and control of the water around the island. The next step, if the port were to flourish, was to devise a way to build needed piers and wharves along the shoreline. Hendrik Hartog, in *Public Property and Private Power*, details the evolving legal status of the city of New York from 1686 to 1870.[10] Two English governors, with the king's permission, gave New York the charters that established the city as a corporate entity and defined the responsibilities of its government, as well as the rights and duties of its citizens.

In 1686, the Dongan Charter recognized New York as a corporate body, "Whereas the city of New-York, is an ancient city within the said province, and the citizens of the said city have anciently been a body politic and corporate." The charter formally referred to the city's government as the "Mayor, Aldermen, and Commonalty of the said Corporate city of New-York."[11] The charter also gave the city all of the "waste land"—the undeveloped land on the island—and specified that the city's jurisdiction extended to "all the rivers, rivulets, coves, creeks, and water-courses, belonging to the same island, as far as the low water mark."[12] The Dongan Charter explicitly granted legal control of the shoreline and the surrounding water—out to the low-water line—to the city of New York.

Commercial activity along the shoreline of the East River continued to increase into the first years of the eighteenth century. As New York's maritime commerce grew, so did the demand for space on the waterfront along the East River. Merchants and shippers pressured the British governors to grant the city the rights to the water surrounding Manhattan Island, in order to stimulate the development of the waterfront.

In 1730, Governor John Montgomerie granted a new charter, which established the city of New York as the local government, independent of the colony of New York. Legal scholars regard the Montgomerie Charter as the foundation for the government of the city. Into the twentieth century (and beyond), the charter "could be seen as the residual source of authority, as part of the 'organic' law of New York City."[13] Hartog describes the structure of the city's government as a combination of public and private powers, a complex political arrangement that echoes into the twenty-first century.

A key part of the Montgomerie Charter granted the city the right to the land, then underwater, extending out from the shoreline for an additional 400 feet beyond the low-water line. The charter defined the underwater land around Manhattan as "water-lots," the equivalent of the undeveloped "waste land" on the island, creating a public asset of immense value. Along the waterfront, the water-lots would provide additional space to build wharves and piers for the ships crowding the harbor. The adjacent waterfront streets would provide access and space for warehouses and business offices for the city's merchants. All of this development required a significant investment, however, and some crucial questions still remained. Would the city of New York underwrite the cost of constructing a commercial waterfront at public expense, as Liverpool did, or would development be transferred to private owners? Would the development and regulation of the waterfront be decided by the city of New York, exercising its inherent political power, or would this be ceded to private interests?

The city of New York required its private citizens, in exchange for the privilege of being part of the "Communality," to pave and clean the streets in front of their property or be penalized. Each year, the Common Council reaffirmed all local laws and published them in the *Minutes of the Common Council*. The laws listed in November 1731 include "A Law for Cleaning the Street Lanes and Alleys of the Said City" and "A Law for Paving the Streets, Lanes, & Alleys within the City of New York."[14]

The Common Council declared that "the former Laws of the City Made for Paving the Streets within the same have been much neglected," and, as a consequence, "the Intercourse of Trade . . . [is] thereby much lessened."[15] Visitors and residents complained constantly about the rutted streets, filled with garbage and turning to mud whenever it rained, creating an impediment to commercial activity. The Common Council placed the burden for paving the streets on the private property owners, stating "that all and Every the Citizens, Freeholders and Inhabitants that dwell in the Respective Streets Lanes & Alleys of this City, or are or shall be in Possession of any Lott or Lotts of Ground fronting any of them shall . . . Well and sufficiently Pave or Cause to be well and sufficiently Paved with good and sufficient Pibble Stones . . . and keep and Maintain the same in good Repair . . . directed by the Alderman and Assistant of each Respective Ward."[16] Thus all expenses for paving the streets and maintaining them in proper order were to be paid by the private property owners, not by the city.

With no municipal garbage system, the streets fronting each house became dumping grounds for commercial as well as private refuse, including human waste. In the mornings, people emptied their chamber pots in front of their homes. The city law for keeping the streets clean required that, on every Friday, the citizens "or [their] servants Rake

and sweep together all the Dirt, Filth and soil lying in the Streets before their Respective dwelling Houses" and then have it carted away and
"thrown into the [East] River."[17]

The street-cleaning law also levied fines for throwing animal or human waste in the streets and asserted that if "any Person or Persons
within this City do Cast, throw or Empty any Tubs of Dung, Close
Stools or Pots of Ordure or Nastiness in any of the Streets of this City,"
they would face a fine of 40 shillings, a substantial amount of money in
the 1730s.[18] The city ordered people to "empty their Ordures into the
River and no where else." From the Dutch settlement through the colonial era and until the late twentieth century, the rivers around Manhattan served as the city's garbage dumps and sewers. Residents of New
York in the 1730s were not unaware of the ecological damage caused by
dumping raw sewage into the rivers. Human waste polluted the water
around the docks, and commentators and merchants complained of the
noxious smells. While the city ordered its citizens to keep the streets
clean and dispose of human waste, it viewed cleaning the streets as a
civic responsibility, not the role of government.

In the North American colonial world, no expectation existed that
any level of government would undertake major public projects to build
roads, wharves, or piers. City officials viewed their primary governmental responsibility as support for the growing commercial activity
of the port. Persistent political pressure came from the shippers and
merchants, whom George Edwards characterizes as "by far the most
powerful" interest groups in the city. The rapidly growing trade in the
Port of New York "brought wealth to many inhabitants. Sons usually
followed fathers in an established mercantile business, so that family
fortunes tended to increase. . . . Through common economic and family
ties were formed combinations potent in politics as well as business."[19]
Above all, the shipowners and merchant princes demanded that the
city facilitate the development of the waterfront.

≈≈ Creating a Commercial Waterfront: Water-Lots

The city of New York controlled one incredibly valuable asset, the water-
lots. It utilized water-lot grants to develop needed maritime facilities,
built by private interests, using private capital, who expected private
profits in return. As Hartog writes, "The property rights of the Mont-

gomerie Charter [water-lots] were granted in pursuit of the goal of creating a major seaport in New York City."[20] The water-lot grants gave private individuals ownership rights to the underwater land along the shoreline out into the East and Hudson Rivers. Typically, the grants identify the city of New York as "the party of the first" and the grantee as "the party of the second." They then define the terms: "do grant, bargain etc., unto said party of the second part, his heirs and assignees forever all that certain water-lot, vacant ground and soil under water to be made-land and gained out of the North [East River] or Hudson River."

First the English kings and then, in the eighteenth century, the city granted these prized water-lots to favored individuals. The beneficiaries of the grants, usually the owners of adjacent shorefront property, "came to realize the significance of the newly acquired grant to the city [and] they became eager to purchase from the city the water lots which lay in front of their own property and thereby acquire that 400-foot territory . . . anticipating the commercial growth of the city."[21] In return, the grantees were obligated to extend the shorefront out into the rivers and construct, at their own expense, a new city street and a bulkhead wharf along the river. Private owners collected the wharfage fees ships paid to moor to the shore, in return for their legal agreement to maintain the bulkhead and wharves, as well as to keep the river along the shoreline dredged. All of the legal requirements of the water-lots had one purpose: to encourage private capital and private interests to build, maintain, and expand the city's port facilities.

For example, Peter Lorillard's grant, between Vestry and Desbroffes (now Desbrosses) Streets, required him to construct a new waterfront street, West Street, along the Hudson River. The new section of West Street was to be 70 feet in width, with "a good and sufficient firm wharf or street" for 380 feet along the Hudson. On this new shore, Lorillard was required to construct a wharf for ships to tie up to and, "at their [Lorillard's] own costs, charges and expenses [to] uphold and keep in good order and repair the whole parts of said streets and wharf."[22] A final clause required that this new portion of West Street be perpetually "for the free and common use and passages of the inhabitants of said city." Lorillard, in turn, could collect all of the wharfage fees from ships using the bulkhead, or wharf, he had constructed.

The real private gain for Lorillard, however, consisted of the two new parcels of made-land along West Street, to be added to the inland prop-

erty he already owned between Vestry and Desbrosses Streets. Lorillard, like all the other water-lot grantees, held title in fee simple (i.e.,
absolute title) to the land created by filling in the waterfront and extending the shoreline out into the river. The city, in turn, had valuable
real estate added to the waterfront, and it obtained the needed extra
wharf space for maritime commerce to flourish.

In some cases, the grants specified an amount the grantees would
pay the city for their water-lot. Fees varied from "one peppercorn" to
the 24 pounds, 11 shillings Carlyle Pollock paid in 1799 for a water-lot
on the Hudson River, between Rector and Carlyle Streets. In return,
Pollock's obligations included the requirement to, "within three years . . .
at his and their own costs and charges and expense, build, erect, and
make . . . a good, sufficient, and firm wharf or street seventy feet in
breadth."[23] This 70-foot-wide street became West Street, the new
shoreline on the western side of Manhattan. Pollock also agreed to fill
in his water-lot and build a pier out into the Hudson River, if requested
to do so by the city.

The Montgomerie Charter left the city with some regulatory power
over the day-to-day operations of the port, however. While the water-
lot grants transferred development of the port to private interests, the
city retained the right to set wharfage fees, that is, the fees ships paid
to tie up to the private docks. The city appointed dockmasters to allo-
cate scarce wharf and pier space to the flood of ships arriving each day.
The city also constructed a number of piers and ferry slips, at public
expense, and leased them to private individuals to operate.

East River Water-Lots

In order to rebuild the waterfront, the newly formed Department of
Docks needed to identify and catalogue all existing grants. In 1871, the
department prepared detailed maps of all the Manhattan water-lots, as
part of their master plan for the waterfront. The mapmakers used the
1776 Ratzer map, drawn by British naval cartographers, as their base
map, which included both high- and low-water lines—marking how far
the high and low tides reached on land—for this section of the East
River. Pearl Street (Queen Street in the colonial era) marked the original high-water line, and Water Street, a short block south, marked the
low-water line. The made-land between Pearl and Water Streets came
from the initial water-lot grants made by the English crown. Later grants

created the new land between Water, Front, and South Streets, moving the shoreline 700 feet out into the East River.

Decades before the Montgomerie Charter, in 1686, King James II granted Peter De Lancey [sic] the city's first water-lot, between Broad and Moore Streets along the East River.[24] De Lancey was a Huguenot who married into the powerful Van Cortlandt family. In March 1722, King George granted a water-lot to Rip Van Dam and "18 others" between Coffee House Slip and Maiden Lane, creating two new city blocks between Water and Front Streets. The early grantees built Water Street and a bulkhead wharf between Old Slip and Maiden Lane. In turn, at their own expense, they filled in their water-lots between Pearl and Water Streets, forming 4 acres of new made-land along this section of the East River. Filling in the water-lots created real estate that belonged to the grantees, not to the city of New York, giving them a benefit of immense value. At the same time, the water-lot grants promoted the construction of a commercial waterfront.

Empowered by the 1730 Montgomerie Charter, the city of New York began to award water-lots. Anthony Rutgers obtained a grant from the Common Council for "water-lots along East River between the Great Dock and Whitehall, the Corner Lot on the East Side of the Street commonly called the Broadway."[25] Rutgers, a wealthy merchant, was among those most well-connected politically. In 1730, he was appointed by Governor Montgomerie to serve as an alderman in the Common Council, the same Common Council that granted him a water-lot.

In the next two decades, the city made only a few more grants, but in the 1750s, the number of grants increased to fifty-seven.[26] On July 10, 1772, four years before the start of the American Revolution, the Common Council approved grants to the east of Rutgers's water lot—extending to Coenties Slip—to seven individuals: Augustus and Frederick Van Cortlandt, Peter Jay, John Vredenburgh, Joris and Henry Remsen, Henry Holland, Waldron Blau, and William Milliner. The grants reflected the preference given to wealthy landowners along the shoreline, as all seven owned houses on Water Street. The growth of the city of New York and the construction of thousands of commercial and residential buildings ensured a steady supply of dirt and rubble to act as fill for the water-lots. Excavations for each new building created hundreds of carts' worth of materials, which were dumped along the East River shoreline.[27]

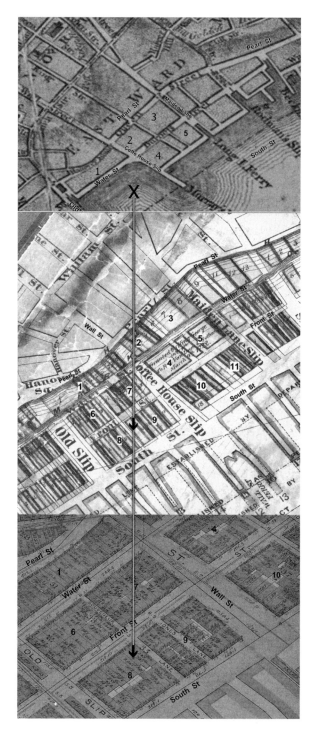

Map 2.2. Water-lot grants on the East River, from Old Slip to Maiden Lane. *Top*: The earliest water-lots (1776), from Pearl Street to Water Street. *Middle*: Additional water-lots (1800–1850s), extending out to South Street. *Bottom*: The same area in 1857, with the water-lots filled in (i.e., made-land) and real estate developed on them, containing counting-houses, warehouses, factories, and tenements. *Source*: Created by Kurt Schlichting. Source GIS layer: historic street maps, New York Public Library, Map Warper.

After the Revolution, the city continued to grant water-lots on into the early nineteenth century. Farther up the East River—between Old Slip, Coffee House Slip (now Wall Street), and Maiden Lane—a bewildering number of grants filled in the waterfront blocks from Pearl Street to South Street (map 2.2). The Department of Docks map identifies 149 water-lot grants. Between Old Slip and Maiden Lane, these water-lots added over 21 acres of made-land to Manhattan Island.

South Street became the new East River shoreline. The final step in the building of the city's maritime infrastructure involved the construction of piers out into the East River. By the time of the Revolution, Water Street (then called Burnels Street) and Front Street defined the harbor frontage, with wharves along the two streets and a number of small piers out into the river. The original grantees of the city blocks filled in their water-lots from Pearl Street out to Water and Front Streets. At the foot of Maiden Lane, the Long Island ferry provided a connection across the river to Brooklyn.

In 1810, the Common Council designated South Street, extending farther out into the East River, as the new waterfront and then awarded fifty-four additional grants. Over time, the water-lot grantees filled in these new lots, creating six additional city blocks. To finance the building of the new South Street piers, the water-lot owners formed partner-

Table 2.1.
Water-lot grants

Location (city block nos. on map)	No. of grants	Acres
Old Slip to Coffee House Slip [Wall St.]		
Pearl St. to Water St. (1)	9	1.5
Water St. to Front St. (6, 7)	18	2.8
Front St. to South St. (8, 9)	18	2.8
Total	45	7.1
Coffee House Slip [Wall St.] to Maiden Lane		
Pearl St. to Water St. (2, 3)	8	4.0
Water St. to Front St. (4, 5)	18	5.1
Front St. to South St. (10, 11)	18	5.3
Total	44	14.4

Source: Water-lot grants, 92–61, box 1, folders 1–4, Department of Ports and Trade, New York City Municipal Archives.

ships and shared the proceeds from the wharfage. They remained jointly responsible for maintenance and improvements. They were also required to keep the water depths between the piers dredged. The city of New York constructed Piers 15 and 16 for the ferry to Montague Street in Brooklyn and leased the operation of the ferry to a private company.

Over the next decades, commercial and residential development followed on this new made-land along the East River. By 1857, the made-land from Pearl to South Streets was filled with maritime-related businesses. Taverns and brothels served the sailors who manned the sailing ships that were tied to the piers in the East River. Tenements crowded Water and Pearl Streets, where longshoremen and their families lived in the crowded immigrant neighborhood along the East River. Tens of thousands of immigrants arrived in New York each year, and they found housing available, even though conditions in many of the tenements were wretched.

All of the five-, six-, and seven-story buildings on the 1857 map were privately owned. Some of the original water-lot grantees developed their property; others sold it to developers. They and their descendants drew great wealth from the teeming waterfront neighborhoods, which were built on made-land along the East River. The city's original intent to develop a maritime infrastructure by granting water-lots had evolved to spawn real estate development along Manhattan's shore.

Stuyvesant Cove

Development of the East River waterfront continued up from the Battery to Corlear's Hook, at Grand Street in the Lower East Side—a distance of over 2 miles. At Fulton Street, the city's fresh fish market prospered, adding to the crush of business on South Street. From north of Corlear's Hook to 14th Street, the city's shipyards expanded and justly earned fame for the quality of the vessels constructed there, including the renowned packet ships. With the ever-increasing number of ships coming to the East River, the demand for space along the shore motivated speculators to press the city to grant more water-lots farther up the river.

Stuyvesant Cove—from 13th Street to 23rd Street along 1st Avenue— offered the last open area on the East River. The river narrowed beyond 23rd Street, and there was not enough space between the Manhattan shoreline and Blackwell's Island (now Roosevelt Island) to build piers. Past the bend at 13th Street, Stuyvesant Cove had once reached as far inland as 1st Avenue. Eric Sanderson's imaginative Mannahatta Project website superimposes his re-creation of the topology of Manhattan Island in 1609 over a Google street map, tracing the expansion of the island into the East River.[28] The cove, once part of the Stuyvesant family's land, which stretched from Bowery Street to the East River, provided a rural retreat where New Yorkers could fish and swim at a time when the city's northern development ended at Canal Street.

Between 1825 and 1846, the city awarded four large water-lot grants for all the land underwater in Stuyvesant Cove. John Flack, an Irish immigrant and shipowner, and Nicholas Gouverneur, the scion of one the oldest and wealthiest Dutch families in New York, received a grant on August 1, 1825, for a major portion of Stuyvesant Cove. Their water-lot extended to the high-water line at 1st Avenue and included all of the underwater land from East 15th Street to East 23rd Street, an enormous area covering almost fifteen square city blocks (map 2.3, #1). The grant required Flack and Gouverneur to build "a good and sufficient firm wharf or street . . . from the Westerly to the Easterly sides," extending out into the river from 15th Street through 22nd Street.[29] From the south to the north, they would construct 1st Avenue, 80 feet in width, and then Avenues A, B, and C. Along the East River, Tompkins Street (later Avenue D) delineated the new shoreline, where the grantees would build a bulkhead for ships to moor alongside. The grant created approximately 58 acres of new made-land, potentially a fortune in real estate once the water-lots were filled.

The grant required Flack and Gouverneur, at their own expense, to extend the city's street grid and, in addition, pay the city a yearly rent. That amount started at $346.13 on May 1, 1826, and increased in each succeeding decade until 1861, when the rent was capped at $1,846 a year. As with all water-lot grants, the grantees were expected to fill in all of the underwater land between the new streets, although the grant did not include a schedule for construction.

For the next nine years, Flack and Gouverneur did not even begin to build any new streets in Stuyvesant Cove. An 1834 D. H. Burr map of the East River shows the cove with no new streets, and the high-water line still at 1st Avenue. What the two partners did was to sell their water-lot to the New York Gaslight Company in February 1848 for $25,933.33. Perhaps they never had any intention of developing the land and instead used their water-lot grant for real estate speculation.

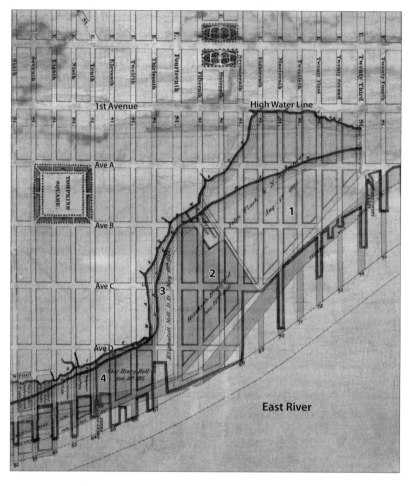

Map 2.3. Water-lot grants at Stuyvesant Cove, on the East River. The grantees and the dates of the grants (indicated by the numbers on the map) were John Flack and Nicholas Gouverneur (1), August 1, 1825; Hezekiah Bradford (2), June 22, 1846; Eliphalett Nott (3), May 13, 1844; and Charles and Henry Hall (4), February 18, 1825. *Source*: Created by Kurt Schlichting. Source GIS layer: 1870 water-lot map, New York Public Library, Map Warper.

The New York Gaslight Company produced manufactured gas to light the city's streets and businesses. All of the city's gas companies employed coal gasification technology to manufacture gas. Natural gas, today delivered by a complex network of underground pipelines from the Midwest, would not arrive in New York for decades. The coal gasification process heated coal at a high temperature in an enclosed boiler to generate gas, which was stored in large tanks and then distributed by pipes under the city streets. Coal gasification produced a toxic byproduct, coal tar, which the gas companies simply dumped into barges and emptied into the harbor. Manufactured gas itself smelled like rotten eggs, because of the sulfur in the coal. Only the poorest residents of the city lived near the gas plants.

The production of manufactured gas required large supplies of coal, so the gas companies were located along the riverfront. The New York Gaslight Company sold their property in Stuyvesant Cove to the Manhattan Gaslight Company. That company built a gas works along 14th Street and then, as the demand increased, expanded their facilities to fill four square blocks from 1st Avenue to the East River, between 20th and 22nd Streets. At 21st Street on the East River, the company constructed a long pier out into the river, where coal barges unloaded. In the 1860s, the area north of 14th Street along the East River became known as the Gashouse District. Two gas storage tanks towered off 1st Avenue and remained there until after World War II, with the construction of Stuyvesant Town. In 1883, all of the independent gas companies merged into the Consolidated Gas Company of New York and eventually became part of Consolidated Edison, or Con Ed, the largest public utility in the country.

Three additional grants in Stuyvesant Cove included one to Eliphalett Nott in 1884, for a narrow water-lot along 14th Street to the East River (map 2.3, #3). Nott's made-land was first a gas works; later, when electricity came to New York, Con Ed converted it to a coal-fired generating station. Disaster struck in September 2012, when Superstorm Sandy inundated New York City. Water surged out of the East River and flooded that Con Edison facility. An enormous explosion followed, lighting up the sky and plunging Lower Manhattan into darkness. Damages exceeded millions of dollars, and Con Ed struggled for over a week to restore the power plant built on the made-land where once the clear waters of Stuyvesant Cove served as a bathing beach.

After World War II, the made-land in the cove became part of Metropolitan Life's massive Stuyvesant Town and Peter Cooper Village housing developments, which, at the time, formed the largest housing project in Manhattan's history. The complex, an 80-acre, tree-lined oasis on the Lower East Side, includes eighty-nine buildings with 8,757 apartments and over 25,000 residents. Most of the complex is built on the made-land that was once Stuyvesant Cove.

The use of water-lot grants to develop the waterfront of Manhattan Island for maritime industry added over 2,000 acres of new land, extending the shoreline out into the surrounding rivers. Stuyvesant Town illustrates the process of large-scale real estate development on the made-land. Yet in an era of climate change, another Superstorm Sandy would threaten the 25,000 people living on that land, highlighting the city's vulnerability.

Hudson River Water-Lots

With the ever-increasing maritime commerce of the port, the East River waterfront could not provide all of the needed pier and wharf space, so the merchant princes turned to the other side of Manhattan Island. The process of creating new land and moving the shoreline outward had begun along the Hudson River before the Revolution. The city made a series of water grants, starting at the Battery and moving north to Rector Street between Greenwich and Washington Streets (see map 1.2). Originally, Greenwich Street fronted the shoreline at high tide. The new water-lots created made-land first between Greenwich and Washington Streets, and then from Washington to West Streets, repeating the pattern of expanding the city outward from the original shore, just as had happened along the East River.

The first grantees owned property along Greenwich Street. On October 28, 1765, the Common Council made grants to Oliver Delancey and Augustus Van Cortlandt. Once again the politically connected triumphed, as early critics had charged.[30] Oliver Delancey was the brother of James Delancey, the governor of the colony of New York from 1753 to 1755. During the American Revolution, Oliver remained a staunch loyalist, became a brigadier general in the British army, and raised a battalion of loyalists to fight with the British. James Delancey's estate, confiscated after the Revolution, became the Lower East Side, with Delancey Street as a main thoroughfare.

In the 1830s, the city designated West Street as the new shoreline along the Hudson River, and its water-lot grants between Washington and West Streets came with the same requirements: the grantees had to fill in the water-lots, construct a bulkhead, build and pave West Street, and construct wharves, all at their own expense. The grantees filled in the water-lots with refuse, and even abandoned ships, but the city's construction boom provided the key, never-ending supply of fill. Excavating a cellar for a typical tenement—25 feet wide by 70 feet deep—produced approximately 520 cubic yards of dirt and rock that had to be carted away. Hundreds of horse-drawn carts moved the rubble to the river's edge, to be used as fill.

As the Port of New York prospered, the demand for pier space along the Hudson River also continued to far outpace the supply. In the 1830s, the city continued to grant water-lots farther up the Hudson, almost to 30th Street. Shipping companies did not want to use piers above West 11th Street, the location of the Gansevoort Market, which supplied the city with food. This market occupied a four-block-square location, where numerous small schooners tied up to the bulkhead daily to deliver fresh food. In 1837, the city persuaded the New York State Legislature to move the shoreline of Manhattan farther out into the Hudson River. The new boundary on paper, 13th Avenue, extended from West 11th to 135th Streets and created the potential for a large number of new water-lots.

Chelsea Water-Lots

In the Chelsea neighborhood north of 14th Street, private waterfront property owners with the right political connections received valuable water-lot grants from the city of New York, some of which were on the expansive scale of those in Stuyvesant Cove. In 1823, Clement C. Moore, the author of "'Twas the Night before Christmas" and the son of the Episcopal bishop of New York, secured a number of water-lot grants to develop the shore of his Chelsea estate, which ran for four city blocks along 10th Avenue, between West 19th and West 24th Streets (map 2.4, section B). At the time of the American Revolution, Manhattan Island's original shoreline ran along 10th Avenue and the Moore estate. Clement Moore's first water-lots, in the 1820s, extended out into the Hudson River for only a few hundred feet. A second set of grants moved the shore farther west, toward 11th Avenue. In 1866, the year Moore died, he obtained another water-lot out to 13th Avenue. Moore's son Benjamin was awarded an enormous water-lot of six city blocks (map 2.4, #6). The Moore family donated the land between West 20th and West 21st Streets to the General Theological Seminary, the seat of the Episcopal Church in New York. The seminary also obtained additional water-lots (map 2.4, #7) and used the property for new buildings. Over time, the Moore water-lots extended the original shoreline of the island

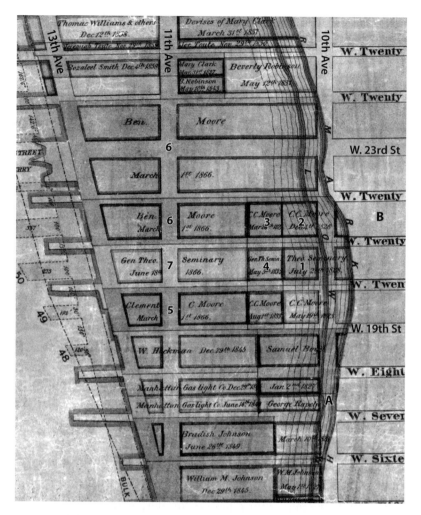

Map 2.4. Chelsea water-lot grants, 1826–1866. The grantees and the dates of the grants (indicated by the numbers on the map) were Clement C. Moore (1), May 19, 1823; Clement C. Moore (2), December 4, 1826; Clement C. Moore (3), March 26, 1827; Clement C. Moore (4), August 1, 1833; Clement C. Moore (5), March 1, 1866; Benjamin Moore (6), March 1, 1866; and General Theological Seminary (7), June 1, 1866. *Source:* Created by Kurt Schlichting. Source GIS layer: historic water-lots map, New York Public Library, Map Warper.

over 1,230 feet out into the Hudson River at 19th Street and 1,530 feet at 24th Street, creating a huge addition of made-land. A hodgepodge of piers built along the new waterfront included the 33rd Street ferry slip to Jersey City.[31]

The Chelsea water-lots not only provided for additional piers from West 14th to West 34th Streets, but the made-land also dramatically expanded the size of the neighborhood, which became filled with private housing, tenements, warehouses, and factories. Longshoremen and immigrants settled in the neighborhood, just as they had along the East River. Beginning with the first water-lot grant in 1686, to Benjamin Moore's Chelsea grants in 1866, and then continuing up the Hudson River to 30th Street and beyond, this process of private—not public—enterprise succeeded in building a maritime infrastructure around the entire southern half of Manhattan Island. Enabled by water-lot grants but financed by private capital, New York's port dominated the maritime commerce of the country (see chapter 3). In the process, the island of Manhattan expanded out into the surrounding rivers and created a new waterfront world. For hundreds of years, the waterfront remained a place of ships, piers, sailors, and longshoremen. Today this waterfront world is a place with upscale housing and spectacular views.

[3]

The Ascendency of the Port of New York

The city of New York and its harbor remained, first and foremost, a place of commerce and trade. At first the waterfront developed to serve a small trading enterprise and, later, the maritime commerce of an emerging United States. All other activities—homes for the wealthy, a place to contemplate the natural beauty of the harbor and its surrounding rivers, and space for recreation—would be relegated to the interior of the island. Nothing could be allowed to interfere with New York's profit-making enterprise, the source of wealth for the city and the country. In 1857, a report to the New York State Senate on "encroachment" in the harbor laid out the importance of using the waterfront for trade: "Maritime nations in every age have sedulously cherished their foreign commerce. Regarding it as an essential means of their prosperity, they have been careful to provide spacious and commodious harbors for its reception . . . [and] expended large sums in improving the natural advantages of their respective ports."[1]

Ninety years later, the Merchant's Association, the Real Estate Owners Association, and the Board of Trade and Transportation all opposed the expansion of Riverside Park, north of 72nd Street, to the edge of the Hudson River. They argued that the continued increase in commerce in the Port of New York would eventually require building new piers as far up the Hudson as 72nd Street and beyond. Maritime use, they argued, took precedence over increased park space.[2]

New York's Rivals for the Atlantic World

Despite the advantages of its superb harbor, the city of New York did not always dominate the trade of the colonies. During the American

Revolution, including the British occupation of the city, its maritime commerce collapsed. The British controlled all shipping in the harbor and allowed only loyalists to trade. Estimates are that during this occupation, the city lost half of its population.

Just days after the British troops landed on Manhattan Island, a major fire broke out on the night of September 21, 1776, which destroyed 25 percent of all buildings in the city. The British accused the "rebels" of arson and placed the city under harsh martial law, which continued throughout the occupation. Neither Baltimore nor Boston nor Philadelphia suffered as much as New York. Many feared that the city would never recover and that one of its competitors would emerge as the new country's major seaport.

Despite that pessimism, New York did rebound, led by a bold group of merchants and shipowners. Exports rose from less than $400,000 in 1783 to almost $3 million by 1793. The voyage of the *Empress of China* in 1784 symbolized the rebirth of the city's maritime commerce. On a cold morning on February 22, the *Empress* departed from an East River pier and set sail for China by way of Cape Horn. Arriving in northern California, the sailors slaughtered seals for their fur, which was of great value in Asia, and then crossed the Pacific Ocean. Fourteen months later, on May 11, 1785, the ship arrived back in New York, setting off a large celebration. The voyage of the *Empress* created a trading link between New York and the Orient. The voyage was estimated to have netted profits in excess of 100 percent for the merchants who risked financing it, including Robert Morris of Revolutionary War fame. Despite the risks, New York merchants sent other ships to follow the *Empress* to China.[3]

After the Revolution, trade with Britain also resumed, despite the lingering bitterness on both sides of the Atlantic Ocean. Above all else, one overriding economic factor linked the new United States to Britain: cotton. By 1800, the Industrial Revolution, which would change the history of both countries and the entire world, was underway in England. A remarkable group of British craftsmen and entrepreneurs transformed the iron industry, built the world's first steam engines, and developed the machinery to mechanize the production of thread and cloth. Soon the river valleys of northwestern England would see the rise of the first textile mills, creating an insatiable demand for raw cotton. With the profits from the cotton textile industry, England rose to become the world's superpower.[4]

While trade across the Atlantic Ocean from New York to Britain expanded, political tensions remained. England soon found itself at war with France, a war that lasted for over a decade. England declared a total embargo on French imports and exports and used its navy to force other countries to cease their trade with France. British and French privateers began to raid Atlantic shipping vessels, including American ships. The British Royal Navy, by far the largest in the world, needed an endless supply of sailors. Among other practices, it began the impressment, or forced recruitment, of seamen from American ships, who were accused of being deserters.

President Thomas Jefferson's ill-advised Embargo Act of 1808 prohibited any exports to foreign countries, which had a devastating impact on New York and all other major ports until the act was repealed a year later. Despite this repeal and the recovery of trade across the Atlantic Ocean, tensions between the United States and Britain continued and eventually led to another war in 1812. New York and its merchants suffered as privateers preyed upon American shipping based in the port. When that war ended in 1815, New Yorkers faced the daunting task of rebuilding the city's maritime economy.

∿ The Black Ball Line Packet Ships

New York Harbor's rise to a position of maritime supremacy began on the transatlantic waterway to Europe. To the east, across 3,000 miles of the North Atlantic, were the major European ports: London and Liverpool in England, and Le Havre on the Normandy coast of France. Albion describes the New York–Liverpool crossing as "the most important of all the world's sea lanes . . . with branches leading to other British or continental ports."[5] This all-important route across the Atlantic Ocean constituted a crucial maritime highway, linking the cotton kingdom in the New World to the Industrial Revolution in the Old World. To secure this waterway, New York shipowners revolutionized the shipping industry, creating something completely new: service on fixed dates, or what Albion calls "square-riggers on schedule."[6]

In ports throughout the world in the early eighteenth century, sailing vessels of all sizes arrived and departed without any specified schedules. Along the East River, ships leaving for Europe, traveling down the coast to the American South, or heading for New England by way of

Long Island Sound waited until they were filled by cargo and passengers. Captains and shipping merchants frequented the Tontine Coffee House, on the corner of Wall and Water Streets, to study both the lists of incoming merchandise for sale and the notices of ships seeking goods and passengers for departure. Some boats advertised a date of departure, but most seldom departed on the advertised date, since the owners saw no reason for their vessels to sail without a full hold and simply postponed the departure. In the meantime, merchants had to wait to load cargo, and passengers had to find a boarding house for temporary accommodations, which cost more money. Transient ships serving New York picked up cargo wherever it could be found and moved from port to port with no set schedule. Merchants could not depend on the transients to deliver cargo to a particular port on any specific date. For example, in 1828, the *Midas* sailed from Salem, Massachusetts, to Portland, Maine, and then on to St. Thomas, Puerto Rico, Mobile, Le Havre in France, and finally to New York.[7] Other ships sailed back and forth on a fixed route, such as New York to Boston or Havana or Mobile or London, but again with no predetermined schedule.

The *Gazette and Commercial Advertiser*, the principal shipping newspaper published in New York in the 1800s, carried advertisements for boats leaving from the waterfront. On Friday, November 15, 1816, the *Gazette* advertised five vessels sailing for Liverpool, including the *Nestor*, "a superior and well known ship . . . built in the city of live oak and locust and coppered last year in Liverpool. Having *now* the principal part of her freight engaged and going on board, she is intended to sail by 25th inst. without fail."[8] The *Nestor* may have sailed by November 25, or perhaps not until December 25. On the same page, two ships advertised service to Dublin, and four to New Orleans. Hundreds of vessels came and left from New York in 1816, but none sailed on a fixed schedule. The same pattern prevailed in Boston, Baltimore, and Philadelphia.

On October 27, 1817, an ad appeared in the *Gazette*, with a woodcut across the top depicting four ships and announcing a "LINE OF AMERICAN PACKETS BETWEEN N. YORK AND LIVERPOOL." A new era began that year, initiating a revolution in ocean transportation on the most important sailing route in the world. Five of New York's prominent shipping merchants—Jeremiah Thompson, Isaac Wright, his son William Wright, Francis Thompson, and Benjamin Marshall—announced that their four ships, the *Amity*, *Courier*, *Pacific*, and *James Monroe*, would begin regularly scheduled service to and from Liverpool each month, departing on a specific day and time. The ad explained that "one of these vessels shall sail from New York on the 5th, and from Liverpool on the 1st of each month," providing scheduled service to Liverpool every month of the year, leaving from Pier 16 on the East River, at the foot of Wall Street. The ships became known as packet ships, because of the packets of mail they carried across the Atlantic Ocean, which was a lucrative source of income (fig. 3.1).[9]

Jeremiah Thompson, a Quaker born in Yorkshire, England, in 1784, arrived in New York in 1801, at the age of 18. Thompson and his partners risked their considerable wealth on their new enterprise, the Black Ball Line, providing the bold leadership that drove the Port of New York to dominance in the Atlantic trade. They maintained a close business relationship with the Liverpool import firm of Crooper, Benson & Company, whose principal owner, James Crooper, also a Quaker, became a leader of the English antislavery movement. Jeremiah Thompson became one of New York's wealthiest merchants and remained very active in the New York Manumission Society. Despite their personal beliefs, both Thompson and Crooper nevertheless built a trans-

Figure 3.1. The Black Ball Line packet ship *Isaac Webb* leaving New York Harbor, bound for Liverpool, England. *Source*: Granger Archives

atlantic partnership inexorably tied to the cotton raised in the American South by millions of slaves.[10]

The Black Ball Line packets sailed in the spring, fall, and summer, as well as throughout the winter months, when the weather in the North Atlantic Ocean was at its worst. Promptly at 10:00 a.m. on January 5, 1818, with snow falling, the *James Monroe* cast off from Pier 16, trimmed its sails, and proceeded down the Upper Bay, through the Narrows, passed Sandy Hook, and set an easterly course for Liverpool, 3,000 miles away. On New Year's Day, the *Courier* left Liverpool bound for New York.

The *James Monroe* landed in England on January 28, after twenty-three days of sailing west to east from New York. The *Courier* faced a much more daunting challenge, crossing the North Atlantic in winter from east to west. Battling against the prevailing winds, she did not reach New York until February, a passage of forty-seven days. By the time the *Courier* arrived in New York, the *Amity* had already departed from New York to Liverpool for February's scheduled sailing. Given the prevailing westerly wind in the North Atlantic, the Black Ball Line ships averaged much faster passages from New York to Liverpool (twenty-four days, or more than three weeks) than the reverse (thirty-eight days, or more than five weeks). The longest averages for crossings by Black Ball ships occurred during the winter months, from December through March.[11]

With the wind pushing a sailing ship from behind, mariners referred to the voyage as sailing "downhill." Ships like the three-masted Black Ballers, with square-rigged sails, faced a much more difficult challenge sailing upwind, with the wind coming from the west, when sailing to New York. The vessels had to tack back and forth, a difficult maneuver for a big sailing ship, especially with the wind blowing hard. To tack, the bow has to be turned across the wind and the sails shifted from one side of the ship to the other, which required great skill and timing with all hands on deck as almost the entire crew manned the maze of lines that controlled the sails. The captain, standing near the sailors who were on the wheel that steered the ship, waited for precisely the right moment to order the tack, while other sailors hauled on their assigned lines. Tacking with a big sea that was running 20- to 30-foot waves, in the middle of a freezing January night in the North Atlantic Ocean, tested the courage and fortitude of the captain and crew.

~~ Rival Packet Lines

The Black Ball Line proved to be an immediate success, with their packets capturing the market for shipping high-value freight and wealthy passengers across the transatlantic waterway to Liverpool. The most important cargo carried by the first Black Ball packets consisted of cotton bales bound for England's expanding cotton textile industry. On their return voyages, the ships brought British textiles and manufactured goods to New York, providing very substantial earnings for the owners.

The Liverpool Customs' bills of entry list the arrival of ships in Liverpool and the cargo they carried. The Black Ball Line's *William Thompson* entered Prince's Dock on July 16, 1825, its hold filled with 887 bales of cotton and 50 barrels of iron ore, with the value of the cotton far exceeding that of the iron ore.[12] The voyage of the *William Thompson* illustrates the "cotton web," which Gene Dattel describes as the first "truly integrated national and international operation, perhaps the first global business. Cotton connected the South with the North, the West, and Europe. Finance, politics, business structure, transportation, distribution, and international relations were all parts of the cotton web. A sophisticated business apparatus developed to service cotton production and the cotton textile industry."[13]

The web linked the cotton plantations in the South to the docks along the East River, the counting-houses in New York City, the Black Ball ships crossing the North Atlantic to Liverpool's Prince's Dock, and the Crooper, Benson & Company import firm. From Liverpool, the web led to the satanic-like mills in Manchester and Birmingham. The voyage back across the Atlantic Ocean brought a flood of textiles, manufactured goods, and British silver and gold. New York shipowners and merchant bankers took their profits, and the remaining specie went farther south, to finance the plantations and slavery. Cotton played the key role in the growth of American maritime commerce, the Port of New York, and the First Industrial Revolution in England. Dattel points out that Karl Marx recognized that "without cotton you have no modern industry." Marx also added, "Without slavery you have no cotton."[14] In the nineteenth century, cotton was the foundation for the wealth, power, and prosperity of both England and the United States.

The Black Ball Line enjoyed a monopoly on fast, scheduled service to Liverpool for five years. Competition came in 1822, with the announcement of the Red Star Line to Liverpool, with four ships. The *Meteor* departed New York on schedule on January 25, a red star painted on its foresail. The Black Ball Line responded immediately and announced the addition of four more ships, as well as departures twice a month instead of just once. With the addition of the Red Star Line, a total of twelve packet ships now provided scheduled service linking New York and Liverpool. In June 1822, another group of bold merchants started the Blue Swallowtail line of packets, and on August 8 the *Robert Fulton* left from the East River, bound for Liverpool. With three lines, packets now left New York on the first, eighth, sixteenth, and twenty-fourth of each month, with corresponding departures from Liverpool. Shippers and passengers could depend on direct, scheduled service from the Port of New York across the Atlantic every week of the month and every month of the year, no matter what the weather.

The New York packet lines also captured the mail business of the Atlantic World and, as international trade grew, their ships provided the most rapid and efficient communication between the United States, England, and continental Europe. New York newspapers became preeminent in the country, based on their access to the most up-to-date news from Britain and western Europe, carried on the packet ships.

No other American port offered comparable service. Packet lines were started in Boston and Baltimore, but with fewer ships and often-missed scheduled departure dates, they could not compete with the Port of New York. In just five years, New York cemented its lead as the premier port in the United States on the most important waterway in the world: across the Atlantic Ocean to Liverpool, the route Albion calls the "Atlantic shuttle."[15] Commerce across the Atlantic Ocean did not just involve Liverpool, however. London became the busiest port in both Britain and the world at the beginning of the nineteenth century. Two packet lines, the Black X Line and Red Swallowtail Line, began service to London in 1824. For thirty years, these two lines sailed between New York and London.

In France, the textile industry, centered in the Normandy region, made Le Havre the country's most important port on the English Channel. Le Havre also traded in French luxury clothing and goods, which became the rage for all fashionable New York women. In 1822, Francis Depau and Isaac Bell announced a packet line to Le Havre, with four ships sailing every other month. The Old Line to Le Havre continued service between New York and France until 1851. Another packet line, the Second Line, began service to Le Havre in 1823. New York became the only port in the country where merchants and passengers could find reliable, scheduled service to England and France, which solidified New York's lead over all other American ports in the Atlantic World.[16]

The success of these packet lines drew the international trade of the United States to the Port of New York. A New York State legislative committee compared the tonnage of imports to New York with those of London in 1830 and 1854. In 1830, the port of London handled 871,204 tons of foreign imports, and New York, just 353,769 tons, 40.6 percent of London's total. By 1854, New York handled 1,919,313 tons of foreign freight, 71.9 percent of London's volume (2,667,823). In just thirty years, New York had become the second-busiest port in the world.[17]

As to overall US trade, between 1830 and 1855, the value of imports and exports grew dramatically. Exports increased from $73.8 million to $275.2 million, or 273 percent (chart 3.1). The value of a single commodity, cotton from the American South, dwarfed all other exports. Imports grew in parallel. With the packet lines well established by 1830,

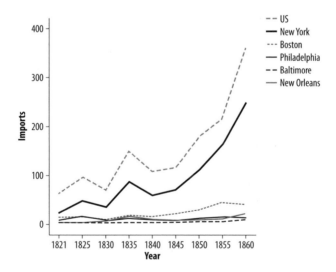

Chart 3.1. US imports at major ports, 1821–1860. The amounts for the imports are in millions of dollars. *Source*: Adapted from Robert Albion, *The Rise of New York Port, 1815–1860* (New York: Charles Scribner's Sons, 1939), appendix 3, 391

over 50 percent of the country's imports were unloaded onto Manhattan's piers.[18]

If the two primary cotton ports, Charleston and New Orleans, are removed, New York's share of the country's foreign trade, especially compared with its northern rival ports, increased dramatically. Over 50 percent of the exports from and two-thirds of the imports to the United States passed across the docks of New York. New York dominated the international maritime commerce of the entire country, while Boston, Baltimore, and Philadelphia fell far behind. The opening of the Erie Canal in 1825 created an inland waterway empire that reached as far west as Chicago and further cemented the Port of New York's ascendency.

≈≈ Tough Ships

The early colonial maps of the New York waterfront identify a number of shipyards along the East River between Wall Street and Corlears Hook. By the early 1800s and the beginning of the packet lines, the demand for larger ships led to the need for expanded shipyards. The waterfront north of Corlears Hook and south of 14th Street became the center of the port's shipbuilding. Just north of where the Williamsburg Bridge now stands, Jacob Westervelt began building packet ships in the 1820s and, over the next three decades, built fifty-five of them, more than any other shipbuilder. The Brown & Bell yard at Houston Street completed two packet ships in 1821, the *William Tell* and the *Orbit*, and, over the next two decades, ten more packets for the Black Ball Line were built in that yard.[19] The third major shipyard was that of William Webb, whose son continued building into the clipper and steamship eras.[20] These shipyards were an integral part of the port's maritime world and became internationally famous for the design and construction of the packet ships that battled the westerly gales in the North Atlantic in winter.

Often a packet ship arrived in New York from Europe bruised and battered. To make the next scheduled sailing, the owners had no choice but to have the ship towed up the East River to one of the shipyards for repair. The skilled shipwrights at the Brown & Bell or the Westervelt yards labored night and day to refit the ship, often at great expense, in time for the next scheduled departure. High shipyard bills became a routine cost for the New York packet lines.[21]

~~~ Tough Men

The fame of the packet ships depended not only on sailing according to schedule, but also on having the shortest passages across the Atlantic Ocean, particularly in the winter months, when they sailed west to New York. Battling the prevailing westerly winds and winter storms, the captains of the packets relentlessly drove their ships day and night. They shortened sail only when absolutely necessary. The fate of the ship depended on the crew, who often were laboring on deck or up the masts in a howling gale, soaked to the bone. Albion describes the sailors who crewed the packet ships as "tough men" who toiled for small wages in extremely perilous conditions. Danger abounded. In the middle of a black night, with a gale raging, snow falling, and ice on the rigging, the risk involved is hard to imagine.

From January through June in 1853, the marine newspapers carried stories of the deaths of eleven seamen on the Liverpool packet ships. On February 14, 1853, on the *Albert Gallatin*, Joshua Remick fell from a yardarm and drowned. As the captain wrote in the ship's log, "It blowing a gale at the time, could not save him." The following month, a log entry for the *Queen of the West* recorded: "12 days from Liverpool, carried away the foreyard [a yardarm on the first mast], foretopmast [top section of mast], lost sails. . . . At the same time Moses Stone, seaman, of New York fell from the masthead and was drown[ed]." With a gale blowing, the *Queen of the West* could not stop and search for Stone, knowing that he was doomed.[22] Later in the year, on June 23, Peter Mulligan, sailing to Liverpool on the *New York*, fell overboard and drowned.

Sailors made their homes in the boarding houses and tenements lining the streets leading down to the East River waterfront. In crowded Ward 1 along the East River, the 1850 census counted over 16,000 people, including wealthy merchants living near the Battery, Irish immigrants, and a population of sailors farther up toward Corlears Hook. In November 1850, the census takers boarded all ships berthed on the East River, to enumerate the officers and seamen living on board. The census included sixty-five ships, with a total of 2,698 crew members, a few officers, and the rest simply listed as seamen, almost all of them born in New York.[23] The 1850 US census counted Peter Mulligan among eighteen seamen aboard a ship berthed on the East River, in New York's

Ward 1. Born in New York, Mulligan, 47, the second-oldest crew member, served with John Haight, 17, the youngest.[24]

Men who made their living at sea often were not literate and were regarded as reprobates. They lived hard lives, many never married, and they died young. Even the Protestant religious reformers, who labored to improve the lives of sailors, viewed them with ambivalence. In 1828, members of the Brick Presbyterian Church founded the American Seamen's Friend Society, to "improve the social, moral and religious condition of seamen . . . to rescue them from sin and to save their souls."[25] This society, along with the Marine Bible Society (1817) and the Society for Promoting the Gospel among Seaman in the Port of New York (1818), built a mariners' church on Roosevelt Street, near the East River.

The Seaman's Friend Society focused on the lives of sailors and their moral redemption. The society published the *Sailor's Magazine*, which could be scathing in its descriptions of the seamen's lives. An 1837 article gave a nod to the hardships they faced but then condemned their wayward life on the waterfront:

> A sailor's life is proverbially a hard one—his toils and sufferings are great—one would naturally suppose that at the end of a voyage he would take care of the small pittance he had hardly earned . . . or in enjoyment of some rational amusement and furnish him with instruction that would prove useful in the after life. . . . They seem to have no thought beyond the present moment—they often seek for pleasure in the indulgence of sensual appetites, at the expense of the moral or intellectual.
>
> The sailor too often divides his time between his boardinghouse, which is often kept by a sharper or pickpocket, a grog shop and a brothel. He associates with the vilest of the vile and sacrifices alternatively at the shrine of intemperance and licentiousness . . . to lure him from the paths of sobriety, virtue and honor.[26]

Hundreds of grog shops and brothels crowded the East River waterfront and nearby streets, and sailors risked their lives on drunken sprees. On a hot August night in 1861, John Williams, a sailor, and a number of his shipmates spent the night drinking in a barroom at 346 Water Street, on the East River. Most of them "became drunk," and a fight ensued. One of the sailors, Augustin Barco, pulled a knife and slashed Williams, "cutting him so severely that his liver was injured and his intestines came out." He died on the barroom floor.[27] A police officer immediately arrested Barco, who was later sentenced to prison.

The 1860 census enumerated a John Williams, a 30-year-old sailor, and his wife Mora, who were residing in a rooming house along the East River, in Ward 4. It was run by Margret Whitson, from Ireland.[28] Unlike many sailors, Williams had a spouse and did not live in one of the notorious seamen's boarding houses, but he was murdered in a bar on Water Street. A day before the murder of Williams, the body of another sailor, Patrick Kearney, was found in the East River, at the foot of Montgomery Street. The coroner ruled that Kearney, age 45 and a native of Ireland, "drowned while intoxicated."[29]

Not only did the sailors fight with one another, but the waterfront toughs, who owned the bars, could also be as vicious. On a Saturday night in December, a year earlier, a police officer came across a sailor named Henry Bentaus, who was lying on the sidewalk on Albany Street. "Believing him to be drunk," the office took Bentaus to a cell in the police station to sleep off his inebriation. The next morning the police found Bentaus dead in the cell and, on close examination, discovered that his skull was fractured. An investigation found that Bentaus and a fellow sailor, Henry Smith, had been drinking in a bar at 11 Albany Street, owned by Philip Collins. An argument over the price of drinks escalated, and Collins "snatched up a club and struck the deceased upon the head."[30] Smith fled for his life, and the police arrested both Collins and his wife, who had participated in the attack.

Like the longshoremen who worked the docks, sailors did not have permanent employment. They were paid by the month, but only while at sea. Once a voyage ended, they received their wages and, battered and bruised, they repaired to the nearest saloon or brothel along the East River to nurse their wounds and tell sea stories.

Seamen faced danger not only on the Atlantic Ocean, but also in the boarding houses, where rapacious owners waited to prey upon them. Boarding house scamps prowled the docks, waiting for a ship to arrive. They then scrambled aboard, with a bottle of rum under their coats, to offer a welcoming toast and steer the weary sailors to a boarding house as soon as they received their pay for the voyage. At the boarding house, food and liquor flowed, and once a seaman had spent his last dime, he lived on credit. In a few weeks he ran up a tidy sum owed to the boarding house, which he could not pay. Usually the bills were wildly padded and, since many sailors could not read, they were victimized. In this situation, a sailor faced just one course of action: to go down to the docks, find a ship ready for sea, sign the papers, and collect a one-month advance on his wages, which the boarding house owner immediately seized. This advance-wage scam persisted for decades, and over 200 boarding houses on the waterfront made tidy sums fleecing the seamen.

A *Sailor's Magazine* editorial railed against the advance-wage system as "always wrong. . . . It is the main trunk of evil. . . . Many are sold under this system like sheep for the shamble."[31] Sometimes a wily sailor would simply disappear and not return to the ship. The landlord instead "sends the man to ship he wishes to get rid of, and the man who owes him the most. . . . He [the landlord] often finds force necessary to get his men upon the ship." In some cases the sailors were kidnapped. In December 1857, a Mrs. Walters went to the police station and charged that her sailor brother, Alexander Chalmers, had been forcibly taken aboard the Liverpool packet *Jacob A. Westervelt*. Chalmers had been living at a notorious boarding house at 76 Roosevelt Street, run by Henry Boorman, where he "was knocked down and cruelly beaten and robbed of all money in his possession, amounting to $14," his advance on wages.[32] The police boarded the ship and searched it but did not find Chalmers. They returned again on Sunday morning. One officer climbed the mast and found Chalmers, bruised and beaten, hidden under a sail. The police then arrested Boorman. Their investigation found that the landlord "has been carrying on this kidnapping business for a long time" and, in the opinion of the prosecutor, "in a manner so flagrant as to secure incarceration in Sing Sing."[33]

Protestant reform ministers and the New York Chamber of Commerce launched a campaign, led by the Seamen's Friend Society, to end the practice of advance wages and have all the sailors' boarding houses registered and inspected by the police. The *New York Times* reported on the reform efforts to end the advance-wage system and instead offer sailors a wage increase of $5 a month if they signed on without advance payment. The reform seemed to be moving forward in 1857, as "one hundred and fifteen New York ship-owners immediately signed an agreement to pay no more advance wages after the first day of July,

1857."[34] The voluntary effort failed, however. Baltimore, Boston, and Philadelphia shipowners refused to go along, and the sailors in New York, against their own self-interest, resisted ending the advance-wage system.

As long as ships crowded the Port of New York, the city remained the temporary home for the tens of thousands of sailors who lived itinerant lives and often died young, racked by rheumatism and consumption from years spent at sea. They might sail one voyage with a good crew, and another with the scrapings of the bars on South Street, with some of the crew members shanghaied drunk and sold by landlords to an outgoing ship. Joseph Conrad captured the poignancy of the deepwater sailors' lives: "A gone shipmate, like any other man, is gone forever and I never met one of them again. But at times the spring-flood of memory sets with force up the dark River of the Nine Bends. . . . Haven't we, together and upon the immortal sea, wrung out a meaning from our sinful lives? Good-bye, brothers! You were a good crowd. As good a crowd as ever fisted with wild cries the beating canvas of a heavy foresail; or tossing aloft, invisible in the night, gave back yell for yell to a westerly gale."[35]

≈≈ Southern Packet Lines

Just as the New York packets dominated the North Atlantic route to Liverpool, London, and Le Havre, the city's shipowners and merchants gained control of the waterway routes along the Atlantic seaboard to the southern cotton ports. Almost at the exact time the Black Ball Line began service to Liverpool, a number of packet lines initiated scheduled service south to the cotton ports: Charleston, Savannah, Mobile, and the all-important New Orleans. Despite the significance of the cotton trade, Albion argues that the role of the southern packet lines has never received the same level of research afforded the transatlantic packet ships. Coastal shipping data is scarce, since the US Customs Service gathered detailed information regarding imports and exports to foreign countries, but little regarding coastal service.[36]

A group of New York shipowners and merchants, with the same drive to succeed as the Liverpool packet-line owners, started the Charleston Shipping Line in 1822, which continued service until 1849. The Charleston Line operated a total of twenty-one ships. The Bulkley shipping family of Southport, Connecticut, a small seaport located in the town of Fairfield on Long Island Sound, started a second Charleston line in 1843, which lasted only until 1850. The Savannah Line began service to Georgia in 1824, while the Mobile Line first sailed in 1826.[37]

Fast, reliable, scheduled sailing led to the success of New York's southern lines, just as it did with the North Atlantic packets. For Liverpool merchants exporting textiles and manufactured goods to the South, it proved to be more reliable and faster to send cargo to New York and then south on a packet ship than to ship it directly from Liverpool to Charleston or New Orleans. Albion's label for New York's control of both the transatlantic and southern marine highways is "enslaving the cotton ports." He states: "The chief exports from New York included cotton, rice and naval stores, grown hundreds of miles to the southward, while the South received most of its foreign imports by way of New York, instead of going to Europe direct, [so] these commodities [especially cotton] have to travel about two hundred miles further. . . . Also there was the added costs of unloading the commodities from one ship and loading them into another at New York."[38]

During the nineteenth century, a very small fraction of American imports from Europe went directly to the South. In 1850, US imports totaled $178 million; only 7.1 percent came through the ports of Charleston and New Orleans, while the Port of New York accounted for 58 percent. By far the most important southern packet service was the one linking the city of New York to New Orleans. Four major packet lines served New Orleans, beginning in 1821 and continuing until the Civil War. The largest, the Holmes Line, operated twenty-eight packet ships, starting with the *Edwin*, the *Chancellor*, the *Lavinia*, and the *Crawford* in 1824, to their last ship, the *Glad Tidings*, in 1858. New Orleans became the major southern cotton port and, by the mid-1840s, was home to more millionaires, whose fortunes were created by the rise of the empire of cotton, than any other city, including New York.[39] The southern packet lines directly tied the Port of New York to the American South and its cotton empire.

With over a hundred packet ships on regular schedules between New York and the southern ports, a great deal of the cotton shipped to Liverpool and Le Havre first went to New York. There it was unloaded, carted from one pier to another, then reloaded onto a Liverpool or Le Havre packet, and carried across the Atlantic Ocean. New York shipping interests profited handsomely. Not only did they own the Atlantic packet ships, but the leading New York merchants also invested in the southern lines. In addition, they owned many of the ships sailing directly from Charleston and New Orleans to Liverpool, further controlling the cotton trade.

Generations of American history books include a map of the "cotton triangle," with the first leg being cotton shipments from the South to Liverpool; the second, textiles and industrial goods from Liverpool to New York; and the third, imports from Europe to the South by way of New York. In many cases, this "triangle" had only two legs: from the South to New York and on to Liverpool; and then back to New York, with the ships' holds filled with imported textiles, manufactured goods, specie, and bills of credit to be settled by the New York banks.

≈ King Cotton

Throughout the entire nineteenth century, the American economy's most valuable product was cotton. All other agricultural and industrial activities paled by comparison. At its heart, the cotton kingdom rested on the labor of an enslaved population of African Americans, as did the economy of the United States. "Cotton had indeed a phenomenal place in the whole American economy; a dominance never since obtained by any single American product of factory and field."[40] In the rise of King Cotton, the port and city of New York played the pivotal role.

The Louisiana Purchase in 1803, and the annexation of the Mississippi Territory in 1804, created a vast inland empire for expansion. President Thomas Jefferson envisioned the settlement of these territories by independent white pioneers, who would labor on their small family farms and provide the bedrock for a democratic polity. Manufacturing would remain in Europe, while, in the American heartland, an agricultural economy built around the small family farm would flourish.

In the South, Jefferson's dream never became reality. On the contrary, an empire of cotton emerged, dependent on "the expansion of black slavery; the racial privilege of the 'empire for liberty' contained within it the seeds of the Cotton Kingdom."[41] Slavery dominated the tobacco, rice, turpentine, and wood economies of the original southern states along the Atlantic seaboard. The first US census, in 1790, counted a slave population of 697,624, with 90 percent (628,107) in Maryland,

Virginia, North Carolina, South Carolina, and Georgia. With the Louisiana Purchase and the expulsion of the Native Americans there, vast new lands in the Mississippi River Valley were opened to the cotton boom. In 1820, the slave population of Alabama, western Georgia, Mississippi, and Louisiana totaled 183,358, and it grew to 385,982—an increase of 110 percent—in just a decade.[42] By 1840, the slave population in the Deep South exceeded the number in the original southern states. Walter Johnson claims that, between 1800 and 1830, over 1 million slaves were moved from the coastal southern states to the Mississippi River Valley, in what he calls a "river of dark dreams." He estimates that one-third left with "their masters as part of intact plantation relocations, the other two-thirds of whom were traded [i.e., sold] through a set of speculations that was quickly formalized into the 'domestic slave trade.'"[43] Owners and slave drivers chained the slaves, bought in Maryland or Virginia, and then mercilessly drove them south to the Mississippi River Valley.

Cotton production increased in the first decades of the nineteenth century and then soared in the three decades before the Civil War. On the new plantations in the Mississippi River Valley, the transported slaves cleared the land, planted cotton, and then harvested it in the sweltering heat of late summer and early fall. During the harvest season, the overseers herded the slaves into the fields at sunup and worked them until sundown. The overseers demanded that each adult slave pick 100 pounds of cotton a day, with their children often toiling beside them.

By 1839, cotton production in the Deep South accounted for over 60 percent of the total cotton production in the United States. In 1860, at the start of the Civil War, three cotton states—Alabama, Mississippi, and Louisiana—grew 78.8 percent of the total. Each bale weighed approximately 400 pounds.[44] In 1810, the number of bales produced in the South totaled 285,800 and rose to a staggering 4,696,553 bales fifty years later.

Long before America's industrial revolution, an agricultural revolution in cotton dominated the country's economy, foreign trade, and the Atlantic World. A wide-spanning cotton web linked the American South, the Port of New York, Liverpool, and northern England, based on black slavery in the South and wage slavery in the textile mills in Manchester.[45] Cotton fully dominated the exports of the United States.[46] By 1830, fifteen years after the War of 1812, cotton totaled over 50 percent of all exports. By comparison, flour, the bounty from the Midwest shipped from the Port of New York, accounted for a modest percentage of all exports: 9.1 percent in 1840, and 4.8 percent in 1860. Cotton production in the South recovered after the Civil War, and once again cotton exceeded the value of all other exports.

New York–based ships provided just one component of New York's involvement. The city's merchant banks provided the financing to grow the cotton and transport it from the plantations to the cotton ports and then across the Atlantic Ocean. A complicated system of credit from the New York merchant bankers allowed the cotton planters, paid once a year, to finance their operations until the cotton reached Liverpool and the mills in Manchester. Payments arrived in New York on the packet ships, and the counting-houses settled accounts. Dattel argues that "cotton propelled New York to commercial dominance" throughout the nineteenth century. New Yorkers played the roles of "the financier, the shipper, the insurer, the factor, the broker, the mill owner . . . [and] the merchant."[47]

Brown Brothers, a firm of New York shippers and bankers, illustrates the complexity of the international cotton business. Alexander Brown, the founder, emigrated from northern Ireland in 1800 and settled in Baltimore, where he opened an importing firm. As his business grew, he sent his sons to open branches elsewhere in the United States and in Liverpool. In 1810, William Brown formed Brown Shipley in Liverpool, which became one of the largest cotton importers in England. His brother James moved to New York and, in 1825, established Brown Brothers & Company, today Brown Brothers Harriman.

In the 1850s, Brown Brothers' estimated worth grew to between $12 and $14 million, and "cotton was one of their main interests."[48] The firm seldom bought cotton in the South for its own account. Doing so was always a risk, given the price fluctuations between paying the planter and then selling the cotton in Liverpool four months later. Rather, Brown Brothers served as middlemen, taking cotton on consignment from the South and then brokering the cotton on its arrival in Liverpool, generating a safe, profitable commission of 2.5 percent. Brown Brothers also arranged for shipping and insurance, adding additional fees.

Their most valuable service involved managing the credit that underpinned the entire cotton web. In 1838, on consignment, the firm handled 1,163,155 bales of cotton exported to Liverpool—15.8 percent of all American cotton—and their commission income totaled $199,738. All of the buying, shipping, and sales involved letters of credit, or bills, which constituted legal tender and could be sold or traded in the southern cotton ports, New York, Liverpool, and London. The trade always involved a discount on the face value of the bills, to assuage the risk involved in buying and selling on credit instead of with hard currency. In 1827, the bills traded by Brown Brothers totaled $132,918.44.[49] With their related firm of Brown Shipley in Liverpool, Brown Brothers in New York moved a vast amount of English and American specie back and forth across the Atlantic Ocean, a service for which the packet ships charged a hefty premium. Between November 1826 and December 31, 1827, Brown Brothers shipped $760,538.98 in specie from Liverpool to New York to settle accounts.

In his prize-winning book, *Empire of Cotton: A Global History*, Sven Beckert contends that merchants, such as Brown Brothers, "were more often independent intermediaries. They specialized not in growing or making, but in moving. At great headwaters such as Liverpool [and New York] they constituted the market; they were the visible hand."[50]

Brown Brothers played another key role in the cotton web, serving directly as what were known as factors in the South. Factors worked out of the southern cotton ports and specialized in providing credit and management services directly to the plantation owners:

> A so-called factor had to first provide the planter with credit to acquire slaves, land and implements. This factor probably drew on London or New York bankers for these resources. Once the cotton had ripened the factor would offer the cotton for sale to exporting merchants in New Orleans [or in other southern ports], who would sell it to importing merchants in Liverpool who would also provide insurance on the bales and organize the shipping. . . . Once in Liverpool, the importing merchant would ask a selling broker to dispose of the cotton. As soon as a buying broker found the cotton to his liking, he would forward it to a manufacturer. . . . The web of cotton consisted of tens of thousands of such ties.[51]

Brown Brothers opened offices in the major southern cotton ports, including New Orleans, in order to provide the services Beckert enumerates. Had the credit system not been available, the expansion of the cotton web into the Deep South would have faltered, and the economic justification for the expansion of slavery would have been decisively weakened. Entries in the books of their New Orleans office record in detail the credit advanced to the plantation owners. For example, this office kept track of all credit advanced to the Bellevue Plantation in Saint James Parish, Louisiana, which was near Lafayette: from the mundane $4 for water buckets, and $70 to repair a cotton gin, all the way up to the salary of the plantation's overseer, one John M. Denneut, which was $800 "for one year to Dec. 31, 1843."[52] Robert Watson, the overseer of the Stamps Plantation, received cash advances of $200 and $400, recorded in March 1844, and then, farther down the page, an entry that simply reads, "Paid A. Peter for two Negroes Maurice & Jean Baptise . . . Cash Dec. 19, 1843, $1,300." Another entry documents the purchase of additional land to expand the cotton slave economy, stating, "Paid Theo Edwards order for A. S. Robertson for the Amt. being lands purchased . . . $3,350."[53]

The New Orleans office arranged to forward the Bellevue cotton, on consignment, from New Orleans to Brown Shipley in Liverpool. In Liverpool, Brown Shipley brokered the sale of "291 Bales Cotton as follows," with a commission of 5.5 percent for Brown Brothers. The cotton sold in Liverpool was paid for in pounds sterling and generated $653.40 in commissions. The New Orleans office also charged the plantation $323.60 for shipping the cotton from New Orleans to Liverpool. The ledger books of the Brown firms include what was listed as "profits" for the Charleston, Galveston, New Orleans, Mobile, and Savannah offices, which totaled $58,691 in 1871 and $62,946 in 1872, each the equivalent of well over $1 million today.[54]

Brown Brothers financed both sides of the Atlantic cotton web and provided the indispensable money, contacts, and expertise to further the most complex commercial relationship of that time period. New York served as the all-important nexus for a century or more. Even the bloody Civil War proved to be merely a temporary interruption.

∾ Liverpool and England's Industrial Revolution

On the other side of the Atlantic World, the Industrial Revolution created an insatiable demand for cotton, the product of American slavery. A remarkable group of mechanics and artisans in northwestern England began a revolution with innovations in iron making, steam engines, and—most importantly for cotton production in the United

States—the mechanization of the textile industry. Without the demand from the English textile mills, there would have been little financial incentive to expand the cotton empire in the Deep South.

Liverpool, Lancashire's window on the Atlantic World, grew to be one of the world's busiest ports, with cotton as its chief import. By 1826, cotton from the United States accounted for 65 percent of all cotton imports to Liverpool from across the Atlantic Ocean. Between 1830 and 1835, it grew to 75 percent and, by 1845, to 80 percent.[55]

The New York–Liverpool packet lines created a virtual marine highway, with cotton ensuring that both port cities flourished. In 1818, the Black Ball Line packets carried 2,004 cotton bales and 6,447 barrels of flour. The following year, that line's freight to Liverpool included 3,210 cotton bales and 3,232 barrels of flour. Cotton had greater value than flour, as freight charges were higher for transporting cotton bales.[56] From the very beginnings of the Black Ball Line and the other Liverpool packets, shipping cotton to Liverpool and textiles back to New York generated a major portion of the profits.

The Port of New York captured a significant share of the cotton shipped across the Atlantic Ocean, shortening the "cotton triangle" to two legs. To illustrate the cotton trade's importance to the Port of New York, in September 1828, fifty-two American ships arrived in Liverpool, transporting a total of 38,995 bales of cotton. Each ship carried an average of 750 bales. The amount shipped directly from New York was 9,828 bales, or 25.2 percent of the total, and that coming directly from New Orleans, 10,034 bales.[57] Thirty years later, in September 1855, twenty-four American ships carried 50,828 bales of cotton to Liverpool, with an average of over 2,000 bales on each ship. New York's share of the total was 22.5 percent, while the ships sailing directly from New Orleans carried 38 percent of the American cotton arriving in Liverpool.

The port of Liverpool, on the north side of the Mersey River, provided the crucial point linking the Deep South to England's textile revolution. The rise of Liverpool as England's great entry port parallels the rise of the Port of New York, but the physical geography of that river and the Liverpool waterfront differed dramatically from New York's.

The Mersey River has the second highest tidal rise and fall in Britain. At low tide, mud flats extend from the shore well out into the river, so piers could not be constructed. For Liverpool's port to prosper, the city of Liverpool, at great public expense, set out to build a marine world completely different from the Manhattan waterfront. The solution was to build wet basins along the shore, which ships could enter at high tide. Then the basin gates would be closed, and the ships moored to waterside quays. As the tide receded, the basins remained full of water.

Liverpool constructed Old Dock, the first of many wet basins, in 1715. By 1821, with the completion of Prince's Dock, the total cost of the basins exceeded £2.6 million, a vast sum for that time.[58] The city borrowed the money for construction and then charged each ship entering the basins, in order to pay back the principal and the interest. The city also managed the docks and used surplus revenue to maintain the basins as a public service. Shipping drawn to Liverpool increased dramatically, just as it did in the Port of New York. On the other side of the North Atlantic waterway, along the Manhattan waterfront, private owners with water-lot grants built additional wharves and piers at a much lower cost.

The building of the Liverpool basins provides an example of public as opposed to private enterprise, such as in the case of the piers and wharves in the Port of New York. To be fair, the amount of capital needed to construct the costly wet basins was beyond the resources of the Liverpool merchants. Liverpool prospered with this public investment in the waterfront. In a history of the rise of the port of Liverpool, published in 1825, Henry Smithers summarizes the impact of the city's investment, listing "the advantages derived from the accommodations afforded shipping by the new docks [wet basins]. They attracted new streams of commerce to the port . . . and amply repaid the investments therein. . . . The docks served to find employ for an industrious and annually increasing population. . . . Increasing wealth . . . [resulted] once [the] 'poor decayed town of Liverpool' reared her head and claimed to rank with the great commercial ports of ancient and modern times . . . and even with London, that emporium of the British empire and of the commerce of the world."[59]

This description mirrors the rise of the Port of New York, with one important distinction: a reliance on private capital and private initiative. The city of New York ceded waterfront development to private interests, something Liverpool never did. Liverpool's public control of the waterfront included regulation of the longshoremen, 100 years before major reforms took hold in New York. In Liverpool the longshoremen no longer had to face the shape-up, but instead worked regularly in a specific basin.

Across 3,000 miles of ocean, New York and Liverpool played complementary roles in the transatlantic waterway empire's cotton web. The rise of the port and the city of New York to commercial dominance depended on the elaborate cotton web, which ran from the slave plantations in the Deep South to New York and, through Liverpool, to the textile mills in Manchester.

[4]

New York's Waterway Empires

With the success of the Liverpool, London, and Le Havre packet lines, the Port of New York dominated maritime commerce crossing the Atlantic Ocean. Ships from all over the world crowded the shoreline, and New York vied with London for the title of busiest port in the world. To the south, the coastal packet lines connected the harbor with the cotton ports in South Carolina, Georgia, Alabama, and Louisiana, cementing the cotton web.

New York Harbor's network of rivers, bays, and waterways offered the city's merchants and shipowners an opportunity to develop an inland waterway empire that tied the port and city to a vast hinterland. The opening of the Erie Canal in 1825 connected "nature's metropolis" to the East Coast. Chicago's rise to commercial dominance in the Midwest depended on a waterway via the Great Lakes, the Erie Canal, and down the Hudson River to the Port of New York. Through New York, Chicago and the upper Midwest traded with the world. While the Black Ball Line packets and the China traders gathered fortunes from distant shores, merchants and shipping interests much closer to home created an inland maritime transportation system as successful as the Atlantic and southern packets. Along Manhattan's waterfront, the Hudson River and Long Island Sound steamboats, ferries, small sloops, and Erie Canal barges competed for limited space with packet ships to Liverpool and majestic clipper ships like the *Flying Cloud*.[1]

≈≈ Inland Waterways

Local waters expanded the city of New York's commercial reach. Like the bold shipowners who sent their packets to Europe, the city's shippers and merchants seized the opportunity to move passengers and freight on a far-reaching complex of inland waterways. These linked the city to a hinterland that extended as far away as Chicago and Canada, as well as to New England and locally to farms in Westchester, Fairfield, Nassau, and Bergen Counties.

The Hudson River, referred to in earlier times as the North River, flows through a picturesque valley and is navigable to just above Albany, 140 miles north of Manhattan.[2] At Albany, the Hudson meets the Mohawk River, which connects to Oneida Lake in central New York State and then to Lake Ontario to the north. To the west of Lake Ontario, the Great Lakes provide waterways to the Midwest, thousands of miles from the Atlantic seaboard. To the north of Albany, Lake George and Lake Champlain lead to the St. Lawrence River and Canada. The East River, a tidal estuary, connects in the Bronx to Long Island Sound. Protected by Long Island, the Sound provides 90 miles of inland passage to New England. Passing south through the Narrows, the harbor joins Raritan Bay and the Raritan River, navigable to New Brunswick, New Jersey, 25 miles from the Delaware River and Philadelphia (map 4.1). No other port on the Atlantic Ocean offered the nexus of inland waterways anywhere near the extent of the Port of New York.

By 1830, over 200,000 people lived on the lower end of Manhattan Island. To feed this multitude, tons of produce, thousands of animals, and an abundance of seafood arrived on the shorefront every day. Oysters came from the expansive oyster beds in the Upper Bay, along the New Jersey shore. Fish arrived at the Fulton Fish Market, on the East River, from the Atlantic Ocean, Long Island Sound, and the Hudson River. Garden farms in New Jersey, Westchester County, Long Island, and Connecticut supplied the city with fresh fruit and vegetables. Wheat and flour arrived not just for export, but also for local consumption.

Manhattan also became a center of the country's industrial revolution. To thrive, companies needed access to raw materials and tons of coal, which provided power. All these materials had to be hauled to the city from the hinterland, which would have been impossible without waterborne transportation. Manufactured goods, in turn, had to be distributed both to markets nearby and throughout the country. The port's inland waterways provided inexpensive transportation decades before the arrival of the railroads.

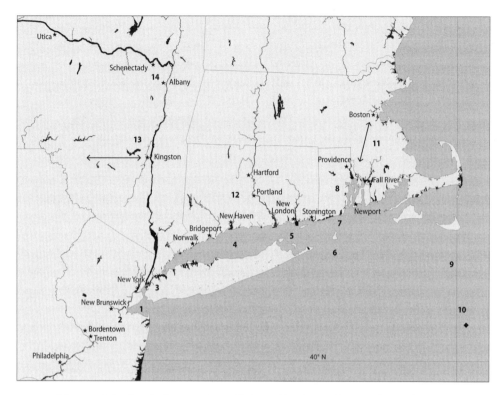

Map 4.1. Inland and other waterways (indicated by the numbers on the map) in New York State and nearby states: Raritan Bay (1); the Raritan River (2); the East River (3); Long Island Sound (4); the Race (5), an entrance to Long Island Sound; Block Island (6), in the Atlantic Ocean; Point Judith, Rhode Island (7); Narragansett Bay, Rhode Island (8); Nantucket Shoals (9), a shallow-water area off the coast of Massachusetts; Nantucket Lightship (10), in the Atlantic Ocean; the 30-mile distance (11) from Providence, Rhode Island, and Fall River, Massachusetts, to Boston; the Connecticut River (12); the 50-mile distance (13) from Kingston, New York, to the Delaware River; and the Hudson River at Albany (14), connecting to the Mohawk River and on to Lake Ontario and the other Great Lakes, as well as to Lake Champlain, the St. Lawrence River, and Canada. *Source*: Created by Kurt Schlichting

Hudson River Lines

In the late eighteenth century, the Hudson River provided the only efficient means of communication to Albany and upstate New York. Travel by sail north to Albany always took longer than the return trip, given the river current. In 1794, the *Albany Gazette* ran a story about a sloop that made the round trip between Albany and New York in a remarkable four days, since a typical trip by sailboat took seven days. With the opening of central New York State to settlement after the American Revolution, travel on the Hudson grew and, as always, speed mattered.

By 1790, ninety sloops operated freight and passenger service on the Hudson River, and in 1800, a group of Albany merchants started a packet line to New York City. Just as the Atlantic and southern packets did, the Albany boats advertised their departures from both Albany and New York on a set day and time. In Manhattan, the Albany packet lines signed long-term leases for pier space, and advertisements directed passengers and freight to a specific location, such as "at the foot of Cortlandt Street" on the Hudson River. The Albany packets provided direct service to the city of New York, but they also added stops at numerous small cities and towns up and down the Hudson River, linking the rise of the city with these smaller communities.

Steam power, which dramatically lowered travel times, came early to the Hudson, with the first voyage of Robert Fulton's commercial steamboat, the *Clermont*, up the river in 1807. An advertisement in the *Albany Gazette* that September listed the *Clermont*'s scheduled departure time from New York City at 6 a.m. on Friday, with an arrival time of 6 p.m. the next evening in Albany, a trip of 36 hours. The *Clermont* promised "good berths and accommodations" for a cost of $7 one way. On October 1, the *New York Evening Post* reported the *Clermont*'s arrival in Albany in 28 hours, with sixty passengers.[3]

The following year, the *Clermont* resumed service as soon as winter ice broke on the Hudson River. Fulton and a number of wealthy investors, including Robert Livingston (of the powerful Livingston family), formed the North River Steamboat Company. Livingston used his political influence to obtain a franchise from the state legislature that granted the company exclusive rights to operate steamboats on *all* of the navigable waters in New York State. The Fulton/Livingston monopoly prohibited any competitors from offering steamboat service on the Hudson River or ferry service to and from Manhattan Island. A political, business, and legal firestorm soon ensued.

Cornelius Vanderbilt, who operated a small sailboat as a passenger and freight ferry between Staten Island and the Battery, decided to enter the steamboat business in New York Harbor. Vanderbilt ignored the Livingston franchise and kept one step ahead of legal attempts to

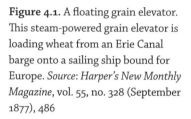

Figure 4.1. A floating grain elevator. This steam-powered grain elevator is loading wheat from an Erie Canal barge onto a sailing ship bound for Europe. *Source*: *Harper's New Monthly Magazine*, vol. 55, no. 328 (September 1877), 486

shut down his business. The ensuing battle eventually reached the US Supreme Court. In 1824, in *Gibbons v. Ogden*, the court ruled that the interstate commerce clause of the US Constitution granted the federal government, not the State of New York, the right to regulate the use of all navigable waterways, including the Hudson River and the waters in the Port of New York.[4] The assertion of federal control over the navigable waterways around Manhattan would play a major role in the development of the Hudson River waterfront.

With the Livingston monopoly canceled, other steamboat lines began service on the Hudson: the Union Line, in 1824; the Hudson River Line, in 1825; and the Morning Line, in 1827. Not to be outdone, Cornelius Vanderbilt started the People's Line in 1834 and, aggressive as always, immediately cut the one-way fare to $1, to try to drive his competitors out of business. With the dramatic increase in steamboats plying the Hudson, passengers were assured of multiple departures each day. By 1845, eighteen steamboats provided service to Albany: on some days, five or six boats passed one another traveling up or down the river.[5] Over the years, the size of the steamboats grew, and the various lines vied to offer the most luxurious accommodations. Some boats carried over 400 passengers, and passengers in first class enjoyed a private stateroom and sumptuous food. Rate wars broke out, and fares fell to as low as 25 cents for a round trip.

Scheduled steamboats also provided passenger and freight service to Kingston, West Point, and other smaller cities on the Hudson River, drawing these communities into the city's economy. Clay deposits along the river in Kingston provided entrepreneurs with the material to man-

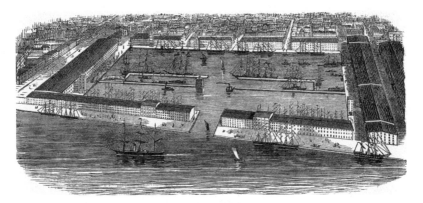

Figure 4.2. The Atlantic Basin, on the Brooklyn waterfront. The enclosed basin had warehouses located directly on the piers. Erie Canal barges and smaller sailing ships unloaded grain and freight, which was safely stored in the warehouses and then loaded onto ships bound for Europe. Ships arriving from foreign ports reversed the process. *Source*: Brooklyn Historical Society

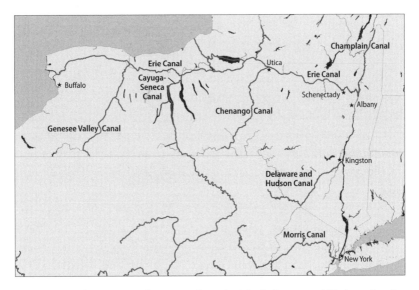

Map 4.2. Canals in New York State and nearby. The Delaware and Hudson Canal and the Morris Canal provided the means to deliver huge amounts of coal and iron ore from Pennsylvania to New York. The Erie Canal linked the Port of New York to the Great Lakes, the Midwest, and Chicago. *Source*: Created by Kurt Schlichting. Source GIS layer: nineteenth-century canals, Jeremy Atack, "Historical Geographic Information Systems (GIS) Database of Nineteenth-Century US Canals," March 2017, my.vanderbilt.edu/jeremyatack/data-downloads/.

ufacture the millions of bricks used to build New York City's homes and factories. Barges carried the bricks down the river to numerous brickyards along the shore. In Cold Spring, across the Hudson from West Point, the West Point Foundry (1817–1911) used local iron and coal to produce artillery weapons and, later, iron pipes for New York City's water system. Farther up the Hudson, in Peekskill, a number of iron works prospered, because of relatively cheap water transportation for their products to New York City.[6]

As with the Atlantic packets, frequent service required the Hudson packet lines to secure permanent pier space on the Manhattan shorefront. Arriving steamboats had to dock as soon as they arrived in order to stay on schedule. By offering a fast and luxurious means of transportation, the Hudson River steamboats continued to operate even after railroads began service between New York and Albany.

The Erie Canal and Canal Fever

New York's inland waterway empire included the all-important Erie Canal, linking the Hudson River with Lake Erie, the Great Lakes, and Chicago. Just as the geography of New York Harbor offered enormous advantages, the unique topography in central New York State provided a crucial benefit. The Appalachian Mountains separate the Atlantic seaboard from the Midwest. From Alabama north to the Canadian border, the Appalachians create a barrier to easy transportation from the East Coast to the Midwest. Only one location offers a "water-level" passage through the mountains: the 300 miles from the Hudson River, up the Mohawk River, and then across an almost-level plain to Buffalo and Lake Erie.

Led by Governor DeWitt Clinton, New York State undertook one of the largest public infrastructure projects in America history: building a canal from Lake Erie to the Hudson River, thereby dramatically expanding New York's inland waterways to the hinterland. When the Erie Canal opened in 1825, wheat and other agricultural goods from the Midwest could be shipped inexpensively through the canal to the piers in Manhattan. Early settlers in central New York State, farming on land confiscated from the Iroquois Indians, took advantage of this low-cost shipping along the Erie Canal to the New York City markets. A bonanza followed. By 1827, an average of over forty canal boats a day passed through Utica, New York, and eastbound freight totaled 185,000 tons.[7]

From the first year the Erie Canal opened, freight and passenger traffic exceeded even the most optimistic estimates. For the next 100 years, the bounty of the Midwest and the farms in New York State floated over the canal to New York Harbor, where thousands of vessels loaded these foodstuffs and departed for Europe. After cotton, wheat and flour were the port's most valuable exports.

Once the canal boats reached Albany, steam tugboats towed these barges down the Hudson River. Thousands of canal boats each year added to the demand for waterfront space. From May to December in 1855, a total of 5,387 canal barges arrived in New York Harbor, transporting 536,324 tons of cargo.[8] Each barge carried an average of 95 tons. Canal boats typically were tied to the bulkhead wharves along South and West Streets to unload. In some cases, the canal barges were moored to Liverpool, London, or Le Havre packets, where a floating grain elevator unloaded wheat directly into their holds (fig. 4.1).

With space at a premium along the Manhattan waterfront, development spread across the East River, along the Brooklyn shore. Two major projects there copied the Liverpool waterfront: the Atlantic and Erie Basins. Rather than constructing piers out into the harbor, the two Brooklyn facilities used breakwaters to create wet basins, where ships and canal boats entered and tied up alongside the wharves. These basins created sheltered pools, with warehouses built on their wharves to safely store cargo, similar to the maritime infrastructure in Liverpool. The Atlantic Basin, built in 1847, enclosed 40 acres, offering space for large sailing vessels and canal barges (fig. 4.2). Samuel Ruggles, a wealthy New York merchant, built additional grain elevators there. The Atlantic Basin elevators dramatically reduced the time and cost of transferring grain from canal barges to ocean-going sailing ships.[9]

The success of the Atlantic Basin inspired a second project to the south, the Erie Basin. Built by Edward Beard in the 1850s, the Erie Basin enclosed a larger area than the Atlantic Basin, and, as traffic on the Erie Canal increased, provided much-needed space. Canal-boat operators wintered over in the Erie Basin when the upstate waterways froze.

"Canal fever" soon followed. Throughout the country, entrepreneurs and state governments—with dreams of economic development and prosperity—set off to build canals. A total of 4,721 miles of canals opened between 1825 and 1876, concentrated in New York, Pennsylvania, and New England (map 4.2). In New York, political pressure led to the construction of additional canals across the state, a number of which generated little traffic and forced the state to use surplus revenue from the Erie Canal to cover these losses. North of Albany, the Champlain Canal connected the Hudson River to Lake Champlain and then to the Richelieu Canal in Canada and the Saint Lawrence River, adding another segment to New York State's inland waterway network, linking agricultural and natural resources to the Port of New York and the Manhattan waterfront.

Two other canals played an especially important role in the industrialization of New York City. The Morris Canal and the Delaware and Hudson Canal both connected the Hudson River with the rich deposits of iron ore and anthracite coal in eastern Pennsylvania. Anthracite coal has a high heat content and fewer impurities than other types of coal. Steam transportation and industrialization in the city created a huge demand for coal, as did the switch from wood to coal for cooking and domestic heating. Coal was also used in the production of manufactured gas for New York City's street lighting. Thousands of coal barges traveled through the canals, supplying the needed fuel. The steamship revolution increased this demand, as did the new factories in Manhattan. Along the waterfront, coal barges were tied up to the bulkhead wharves, where the lowest-paid dockworkers unloaded the canal boats, which was a hard, dirty job. Across South and West Streets, numerous yards stored the coal for distribution, and the horse-drawn coal wagons added even more traffic to the already crowded waterfront.

Long Island Sound Steamers

To the east of Manhattan Island, Long Island extends for over 100 miles to Montauk Point and forms the sheltered Long Island Sound. From Throgs Neck in the Bronx, where the East River meets the Sound, the inland waterway offers 90 nautical miles of protected waters to the Race, the passage that marks the eastern entrance to the Sound (see map 4.1, #5). Both the shoreline of Connecticut and the northern shore of Long Island include numerous ports, where local farmers shipped produce and livestock down the Sound to New York City. While three-masted sailing ships crossed the North Atlantic, small schooners (with two masts) and sloops (with one mast) used the Sound as a protected inland passage. Long after the arrival of the steamboats, thousands of schooners and sloops continued to provide low-cost shipping to and from the New York waterfront. Rapid advances in steam technology would dramatically improve travel time on New York's inland waterways, just as steam power had on the North Atlantic Ocean.[10]

In 1822, the New York and Rhode Island Steamboat Company began service on Long Island Sound, going from New York City to Providence, Rhode Island, with connections to Boston by stagecoach. The inland waterway enabled steamboats and passengers to avoid the long outside ocean voyage from New York City to Boston, traveling around the Nantucket Shoals, which stretch 40 miles south of Nantucket Island (see map 4.1, #9). An outside voyage to Boston required a sailing ship to travel over 300 nautical miles out into the Atlantic Ocean, around the Nantucket Shoals, north along Cape Cod, and across Massachusetts Bay. A trip on the steamer included 90 miles in Long Island Sound; just 30 miles on the ocean, from the Race to Point Judith, Rhode Island; then 6 miles north into Narragansett Bay; and a protected run up to Providence, Rhode Island—a total of 160 nautical miles. At Providence, stagecoaches provided connections to Boston, 30 miles to the north (see map 4.1, #4, #5, #7, #8, and #11). The New York and Rhode Island steamships made one round trip a week, leaving from the East River. The voyage averaged 25 hours one way and cost $10. Thus travelers could reach Boston by steamboat and stagecoach in two days, as opposed to about a week by stagecoach directly from New York City.

The New York and Rhode Island line added new boats in the following years, each larger and with more powerful engines, reducing the average time for the run to Providence to 16 hours. Competition arrived in 1830, when the New York and Boston Steamboat Company and the Providence Steamboat Company both began service. Cornelius Vanderbilt, after great success with his People's Line between New York City and Albany, joined the competition and ordered a new state-of-the-art steamboat in 1834. On its first trip, the Lexington arrived in Providence in 12.5 hours. By the 1830s, however, rail travel began to undercut the steamboats. The Boston and Providence Railroad allowed travelers to leave New York City in the evening and arrive in Boston the following noon.

Vanderbilt's line and the other Long Island Sound steamboat companies then began to schedule a stop in Newport, Rhode Island, at the entrance to Narragansett Bay. After the Revolutionary War, Newport—

once the fifth most populous place in the colonial world—suffered a dramatic decline in its maritime commerce and fell far behind Boston and New York, its former close rivals. Newport's population stagnated, and only the arrival of the US Navy and the construction of massive Fort Adams, on the coast, kept the economy alive. The coming of the Long Island Sound steamboats changed Newport's fortunes.

This locale, one of the most harmonious meetings of land and sea on the coast of North America, is blessed by the moderating warmth of the Gulf Stream, so Newport on Aquidneck Island enjoys mild winters. Newport's harbor seldom freezes, which attracted the US Navy. Once the Long Island Sound steamers provided quick, scheduled service from New York, wealthy tourists followed, and Newport joined Saratoga Springs, in New York State, as two of the country's first resorts.[11] Newport's tourists initially boarded with local families, until a number of hotels opened in the 1840s, including the famous Ocean House on Bellevue Avenue in 1844. With overnight connections to New York City, Newport soon drew wealthy New Yorkers for the summer months. Southern plantation owners and cotton merchants, along with their families, fled to New York during the summer months to escape the sweltering heat and related diseases back home. From New York City, many traveled on the Sound to Newport for the summer, creating strong social connections between Charleston, Savannah, New York, and Newport. These social ties were a direct consequence of New York's waterway empire.[12]

After the Civil War, Newport attracted the country's wealthiest families, many with new wealth that was created by the Industrial Revolution. Robber barons, including the grandsons of Cornelius Vanderbilt, built sumptuous mansions along Bellevue Avenue. The Gilded Age flourished on a nearly unimaginable scale each summer season in these Newport "cottages." An army of servants catered to every need of the privileged few. Many of these servants were immigrants, who had traveled from Ireland to New York aboard the Black Ball Line packets. Their social web connected Ireland with the Port of New York and then Newport, becoming another byproduct of an Atlantic World.

The southern packet lines to New York City and the Long Island Sound steamers drew Rhode Island into the cotton web. North of Providence, in Pawtucket, Rhode Island, Samuel Slater had opened the first successful cotton mill in the United States in 1791, on the west bank of the Blackstone River, marking the beginning of New England's cotton textile industry. By 1812, Pawtucket had twenty-four cotton mills, with 20,000 spindles. To the north, many other mills prospered in the Blackstone River valley. Here, too, the raw material came from the Deep South and the slave plantations.

The cotton bales were shipped to New York on the southern packets, unloaded on the piers along the Manhattan waterfront, and carted to the wharves, where the Long Island Sound steamers departed for Rhode Island. While these steamboats had a growing reputation for luxurious accommodations, profits also accrued from transporting cotton bales to Rhode Island. By 1860 and the start of the Civil War, Rhode Island had 153 cotton mills and employed 13,477 men, women, and children—41.5 percent of the total manufacturing employment in the state. The mills used 41,614,797 pounds of cotton, or over 1 million bales.[13] Revenues from the production of textiles at that point totaled $38 million, versus $21 million a decade earlier—an increase of 81 percent between 1850 and 1860.

The cotton web ensnared both New England and Liverpool, with the Port of New York as the crucial hub. As steamboat service on Long Island Sound prospered and expanded, the cotton web spun through Massachusetts and New Hampshire. The famous Fall River Line started steamboat service from New York to Massachusetts in 1847 and continued its operations for almost a century, to 1937. The Quequechan River flows through the city of Fall River to Narragansett Bay. It provided water power for the city's first cotton mills, built in the 1810s. By 1850, Fall River had become the leading textile-manufacturing center in the United States. Frequent, scheduled steamers afforded cheap transportation for cotton bales and, later, as more mills opened, for the mountains of coal used by the steam-powered mills.

The Fall River Line became synonymous with fast and luxurious service between New York and Narragansett Bay.[14] The line launched the *Pilgrim* in 1893. At 370 feet in length, this steamer could accommodate 1,200 passengers and complete the 176-mile trip between New York, Newport, and Fall River in 8 hours, at an average speed of 22 miles an hour! Every day, passengers and freight on the Fall River Line arrived at Pier 28 on the Hudson River, at Murray Street. In the busy summer months, the line ran two steamboats: one during the day, and a second at night. The latter departed promptly at 6 p.m. and arrived in Newport just after midnight.

The Long Island Sound lines competed fiercely for both passengers and freight, often staging rate wars and reducing the price of a one-way passage to Providence to $1. As a consequence, the cost of transportation between New York and New England decreased, steadily knitting the New England hinterland to New York. Farmers, merchants, and manufacturers could send their products cheaply and efficiently to the Port of New York for either export, reshipment along the coast, or local consumption. Textile manufacturers in Rhode Island and Massachusetts found it cheaper and more convenient to ship their goods to New York than to Boston, just 30 or 40 miles away. In return, cotton bales unloaded on the piers in New York arrived in Fall River or Providence significantly faster than if they were carried by ship from a southern cotton port directly to Boston.

≈ Ferry Service

Some of Manhattan Island's waterways extended for hundreds of miles, and others for much shorter distances: across the East River to Brooklyn and Queens, and across the Hudson River to New Jersey. Manhattan, like any populated island, had to find a way to be connected to the mainland. In 1661, the Dutch governor of the region awarded an early franchise to William Jansen to operate a ferry across the Hudson River to Communipaw, located today within Jersey City. In 1686, the British, in the Dongan Charter, granted the city of New York control over ferry service across both the Hudson and East Rivers.[15]

Periaugers—sail-powered, two-masted, flat-bottomed boats of Dutch design—served as the first ferries. This type of boat, about 20 feet long, could be manned by one sailor and rowed, if necessary, when the wind died. Cornelius Vanderbilt started his rise to great fortune with money borrowed from his parents, which he used to operate a periauger between Staten Island and the Battery in 1810. The first steam-powered ferries began service in 1812, between Cortlandt Street in Lower Manhattan, across the Hudson River, to Powles Hook in Jersey City. In 1814, additional ferries ran from Fulton Street, on the East River, to Brooklyn.[16]

Steam-powered ferry service on the East River accelerated change in the New York City area, strengthening the commercial and social ties between Manhattan Island and Long Island. By the 1830s, parts of

Brooklyn emerged as residential suburbs of Manhattan for the well-to-do, and the world of the commuter began. Successful businessmen built brownstone homes on the quiet streets of Brooklyn Heights or Cobble Hill and rode a horse-drawn streetcar to the waterfront ferries. After a short trip across the East River, they arrived at the counting-houses lining South Street or walked a few blocks to Wall Street, which was rapidly becoming the center of the country's financial markets.

By the turn of the twentieth century, forty-two ferry lines connected Manhattan with Brooklyn, Queens, Staten Island, and New Jersey.[17] Hundreds of thousands of passengers rode on them each day, and tons of freight crossed on the ferries, carried from the waterfront by horse-drawn drays. In 1866, a total of almost 80 million people used the ferries. At the busiest times of the day, two and sometimes three ferryboats ran on one route, leaving the dock every 10 minutes.[18] The tremendous growth of the ferry business created a metropolitan region. At its heart, the Manhattan waterfront served as the lynchpin: all but one or two of the ferry lines terminated on the shore of the city.[19]

Pressure for room along the Manhattan waterfront escalated with the ever-growing number of ocean-going and coastal ships, steamboats serving the inland waterways, and ferries. The ferries needed dedicated space: one or two specially designed slips for the ferries to drive into; and a ferry building, with space for ticket offices, waiting rooms, and freight storage. No other shipping in the harbor could use either the ferry slips or the shoreline occupied by the often-elaborate ferry terminals. Each ferry building served as a focal point, filled with constant activity. The city's first rapid transit system, the horse-drawn street railways, ran their tracks down to the riverfront to provide service for ferry passengers. On the main deck, ferries carried drays loaded with freight, as well as private carriages. Both passengers and freight crowded the ferries day and night, adding to the continuous jam of traffic along West and South Streets.

Construction of the Brooklyn Bridge, the longest bridge in the world when it opened in 1883, furthered the growth of Brooklyn neighborhoods filled with the working-class housing.[20] Brooklyn's population grew from 566,663 in 1880 to over 800,000 a decade later, an increase of 42 percent.[21] In 1898, the consolidation of Manhattan and Brooklyn with the boroughs (counties) of the Bronx, Queens, and Staten Island created what is now known as greater New York City. Part of the argument for this consolidation involved the fragmented control of the waterfront.[22] After 1898, the city's Department of Docks took control of the waterfront in Brooklyn, Queens, and Staten Island (see chapter 7).

~~ New York's Maritime Predominance

By the outbreak of the Civil War, the relentless energy and drive of New York City's merchants, bankers, and maritime businessmen had created a waterway empire that extended across the Atlantic Ocean. New York packets dominated the Atlantic World, and coastal packets linked the city to the cotton kingdom in the South (table 4.1).[23] No other seaport in the United States achieved the success of the Port of New York.

Table 4.1.
New York packet service, 1845

Packet lines	Number	Sailings
North Atlantic		
Liverpool	21 ships	1, 7, 13, 19, 25 of each month
London	12 ships	1, 10, 20 of each month
Le Havre	16 ships	8, 16, 24 of each month
Total monthly sailings		11
Other foreign ports		
St. Johns		once a week
Havana and Puerto Rico		twice a month
Cartagena, Matanzas, and Vera Cruz		once a month
Antwerp, Bristol, and Hamburg		once a month
Coastal, to southern United States		
New Orleans	5 lines	weekly
Richmond and Savannah	3 lines each	weekly
Alexandria, Baltimore, and Charleston	2 lines each	weekly
Apalachicola and Norfolk	2 lines each	weekly
Fredericksburg and Pensacola	1 line each	weekly
Total weekly sailings		19

Source: O. L. Holley, ed., *The New-York State Register for 1845; Containing an Almanac for 1845–6* (New York: J. Disturnell, 1845), 255–258.

The city's inland waterway empire grew in parallel. Steamboats raced to New England through Long Island Sound and up the Hudson River to Albany. The Erie Canal led to the Great Lakes and on to Chicago. In New York Harbor, a fleet of ferries linked Manhattan Island to the mainland and its burgeoning metropolitan region. New York drew the commerce and the growing bounty of the Midwest, as well as cotton from the South, to the piers on the Manhattan shoreline. Freight and passenger lines connected the city to upstate New York. Food, coal, timber, bricks, and brownstone arrived from New Jersey, Long Island, and Connecticut, carried aboard small sailing vessels and wooden barges (table 4.2).

The men behind the shipping revolution followed a common pattern of technological innovation. Samuel Thompson's first packets measured 100 feet in length, with a small cargo capacity. As business flourished, the packet lines ordered increasingly larger vessels, built in the East River boatyards, of the best materials, by skilled craftsmen. The captains spurred their ships across the Atlantic Ocean in all kinds of weather and reduced the time of passage from months to weeks.

The age of steam revolutionized the maritime world, on both the ocean and the inland waterways. Shipowners competed for speed up the Hudson River, down Long Island Sound, or across the Atlantic Ocean. More-powerful steam engines—as well as the switch from paddle-wheel steamboats to those with screw propellers, and from wooden to iron hulls—led to faster times. Relentless competition, driven by the likes of Cornelius Vanderbilt, and forty years of technological innovation dramatically reduced the cost of transportation. The port and city of New York prospered on a nearly unimaginable scale.

〰 Transatlantic Passengers as Immigrants to New York

While the cotton web created the Atlantic World connection between New York and Liverpool, with extensions to Le Havre and London, the packet lines carried passengers as well. Only the wealthy could afford the fare of $120 in the 1820s, with food and wine included. On September 6, 1822, the Black Ball Line packet *William Thompson* arrived in New York from Liverpool with eleven passengers. The ship's manifest included Stephen Pier, age 39; Walter Livingston, age 23; Thomas Bachley, age 70—all New York residents—and Robert Hunt, age 20, from Charleston. Each of them was listed as a "gentleman." Four other US citizens, all

Table 4.2.

New York inland steamboats and canal boats, 1845

Type of boat	Number	Scheduled service	Departure from
Steamboats			
Long Island Sound			
Stonington, Newport, and Providence	5 boats	daily	Battery
Norwich, Hartford, and Bridgeport	3 boats each	daily	Hudson River, Pier 1
New Haven	2 boats	daily	Hudson River, Pier 1
New Jersey			
South Amboy	2 boats	daily	
Elizabethport and Newark Bay	2 boats	4 times/ day	Hudson River, Pier 1
Newark and New Brunswick	1 boat	2 times/ day	Hudson River, Barclay St.
Hudson River			
Troy and Albany	12 boats	daily	
Hudson River ports	24 boats		
From Buffalo to			
Chicago, Detroit, and Green Bay	24 boats		
Toledo and Canada	8 boats		
Canal packets			
Passengers			
Erie: Schenectady, Utica, Syracuse, Rochester, and Buffalo	2 lines	2 times/ day	
Oswego: Syracuse to Oswego	1 line	daily	
Cayuga: Montezuma to Geneva	1 line	daily	
Genesee: Rochester to Dansville	1 line	daily	
Champlain: Troy to Lake Champlain	1 line	daily	
Freight lines			
New York, Albany, and Buffalo	11 lines		
Finger Lakes	8 lines		
Troy, Buffalo, Lake Erie, and Detroit	9 lines		

Source: O. L. Holley, ed., *The New-York State Register for 1845; Containing an Almanac for 1845–6* (New York: J. Disturnell, 1845), 255–258.

merchants, accompanied Charles Mathias, age 45, entered on the manifest as a "comedian" from London, who arrived in New York to perform on stage.

On the winter crossings to New York, the packets carried fewer passengers, given the gales and freezing weather often encountered. The *William Thompson*'s passenger list for its January 29, 1824, arrival in New York contained the names of just five hardy souls, including Sir Rupert George, age 27, who traveled with his manservant, George Priest; and Francis Thompson Jr., a 25-year-old merchant from New York whose father, along with his Quaker partners, had founded the Black Ball Line. New York merchants customarily brought their sons and sons-in-law into the family firms. In this fashion, several generations of the Thompsons ran their expanding shipping empire.

During the nineteenth century—the century of immigration—the Port of New York also served as the primary port of entry for millions of Europeans coming to America. The success of the packet lines on the most established route across the North Atlantic drew the immigrant trade to New York. Between 1820 and 1840, the influx of immigrants remained small. In the 1840s, however, a potato famine in Ireland dramatically increased the number of people from that country emigrating to the United States. German immigrants fleeing economic and political unrest also arrived in large numbers.[24]

The packet lines from Liverpool, Le Havre, and Bremen (in Germany) responded by offering passage to America in the tween deck: the space between the deck and the lower hold. This deck, only 8 or 10 feet in height, had no direct light or ventilation. In good weather the cargo hatches remained open, providing some fresh air and light. These steerage passengers climbed up onto the deck in good weather to cook their own food and use the heads (toilets in the bow of the ship) for sanitation. When the weather deteriorated, the sailors closed the hatches and made them watertight, confining all steerage passengers below. In the winter, storms often lasted for days and weeks, and the hatches remained battened down, creating a hell in the tween deck.

The potato blight set off a disaster of monumental proportions, changing the history of both Ireland and the United States.[25] An estimated 1.4 million Irish—often referred to as famine Irish—left their homeland to relocate abroad, with the United States being the destination of many. A National Archives and Records Administration (NARA) database, the "Famine Irish Passenger Record Data File," lists the names of 604,596 immigrants who arrived in New York Harbor between 1846 and 1851, an average of over 100,000 a year.[26] In New York and other port cities, the Irish arrived in large numbers, creating distinctive neighborhoods, such as Five Points (see map 5.1). In New York, the majority, impoverished and destitute, settled along the Manhattan waterfront.[27] Willing to do the hardest and most dangerous work for the lowest wages, they provided the manual labor needed in the port and the city's factories. The daughters of the Irish diaspora worked as servants for the wealthy, and "Bridget" became the synonym for Irish female servants in New York during the century of immigration.

The famine Irish landed at the piers along the East River. No documents of any kind were required. The vessel's officers simply had to bring a ship's manifest to the customs office, listing the names of the arriving passengers. The State of New York did not establish the first immigrant processing center, Castle Garden in the Battery, until 1855. The US government opened Ellis Island in 1892, almost three decades later, when the federal government assumed the responsibility for immigration processing.

The numbers for one day in one month illustrate the magnitude of the flood of people coming through the Port of New York. On July 25, 1848, the Black Ball Line packet *Columbus*, from Liverpool, sailed up the harbor and moored at Pier 28 on the East River, between Dover and Roosevelt Streets, with 503 passengers aboard. It is nearly impossible to imagine how these immigrants, in addition to the crew members on the ship, managed to endure a passage of over a month on a ship just 138 feet in length, with a beam, or width, of 36 feet.[28] Henry Hudson's *Half Moon*, at 85 feet long, had a crew of only about twenty. While few immigrants left a record of their experience, Herman Melville's autobiographical character, Redburn, describes one such arrival.[29] His ship, the *Highlander*, was filled with immigrants who endured a slow voyage across the Atlantic Ocean. Hundreds of them were crammed into the tween decks. Flimsy wooden bunks, hastily assembled for the voyage, provided a minimum of privacy. The immigrants brought their own bedding aboard. To avoid quarantine, the captain of the *Highlander* ordered Redburn and the rest of the crew to clean out the tween decks and, except for what the passengers could hold onto, throw everything overboard. The debris littered the ocean. The crew then tore down the bunks and threw the wood into the sea. As a final measure, the crew doused the space with carbolic acid, to kill the smell left by hundreds of people crammed together for weeks with little ventilation and limited water in which to wash themselves.

The ship's manifest for the *Columbus* lists most of the 503 passengers' place of birth and "last place of settlement" as Ireland. Bridget Barry, age 50, came with her three daughters—Ann, 18; Barbara, 15; and Hannah, 11—and son Pat, 10. There is no record of a husband, or whether any of her family or friends who were already in New York met the *Columbus*. Coming to America provided most immigrants with a challenge and hardship, and a widow with four children faced even steeper odds.

Three other immigrant ships arrived in New York on the same day as the *Columbus*: the *Alice Wilson* from Liverpool, with 333 passengers; the *Juno*, also from Liverpool, with 256 people aboard; and the *Charlotte Harrison*, from Greenock, near Glasgow, with 155 Scottish immigrants. Even during the height of the arrival of the famine Irish, people came to New York from other parts of the British Isles and Europe. For the month of July 1848, the NARA database lists seventy-six immigrant ships arriving, with a total of 11,532 passengers, for an average of 443 people each day.[30] All seventy-six vessels found dock space along South Street, adding to the waterfront chaos and the crowding in the boarding houses and tenements on the nearby streets.

Ships' manifests for the century of immigration do not provide uniform information. Most list a passenger's name, age, place of origin, and occupation. The *Juno*'s manifest contained additional data, including the immigrant's planned destination in America. Almost all of its passengers declared New York (i.e., Manhattan Island) as their final destination. A few opted to move on to Boston and Detroit, and nineteen were headed to Illinois. Hundreds of thousands of immigrants who arrived in the Port of New York stayed in the city, often mere blocks from the piers, creating an Irish waterfront.

The packet lines did not depend on ferrying immigrants from Liverpool to New York. Their primary business remained carrying cotton and other high-value cargo to Liverpool and, on the return voyage, bringing back textiles, manufactured goods, and specie. Nonetheless, during the famine years, the packet lines also carried a total of 115,210 immigrants from Liverpool, providing their owners with a steady source of profits.[31]

Sir Rupert George paid a premium to cross from Liverpool in relative comfort on the *William Thompson* in 1824; the famine Irish paid about $20 two decades later for no comfort whatsoever.

Two of the Black Ball Line's ships, the *Columbus* and the *Manhattan*, made four round trips between Liverpool and New York in 1849 and 1851, a testament to the hard-driving captains and tough crews. The *Manhattan*'s first arrival, on January 6, 1851, with 561 passengers aboard, meant that this ship departed from Liverpool in late November and spent the month of December on the North Atlantic. The famine Irish endured a miserable time in the tween deck. With gales raging, they could neither cook hot food nor empty the chamber pots.

Even after the worst of the famine years, immigrants from Ireland and Great Britain continued to cross to New York from Liverpool in large numbers. The packets provided the fastest passage, but other freight ships carrying cotton to Liverpool returned to New York with significant numbers of passengers. Between 1846 and 1851, the Black Ball Line packets had carried a total of 36,847 famine Irish.[32] In September 1855, twenty-four American ships arrived in Liverpool: eight from New York, eight from New Orleans, six from Mobile, and one each from Galveston, Texas, and Alexandria, Virginia. The New York vessels delivered 7,720 bales of cotton, brought to that city by the southern packets, illustrating the two-legged cotton triangle.

Once ships had arrived in the Liverpool dock basins, longshoremen worked feverishly to unload them, so the vessels could start back to New York as quickly as possible. On the day of departure, hundreds of emigrants queued up to board, filled with trepidation. Many realized that they would never return—never see their families again. All faced the challenge of building new lives across the Atlantic Ocean. The seven New York vessels came back to the city carrying just over 2,500 passengers, including 2,163 immigrants in the tween decks.[33] On the *Isaac Webb* (see fig. 3.1), the 13 cabin passengers paid $100 each. The 411 immigrants traveling below deck, at $20 each for adults and $10 per child, generated much more revenue than the first class passengers.[34]

For the entire nineteenth century, New York City had the highest percentage of foreign-born of any place in the country. In the middle of the century, tens of thousands of Irish immigrants came to New York, as did thousands of Germans leaving the various states and principalities destined to become Germany in 1871. American immigration has always been driven by a combination of push and pull factors. The potato famine left millions in Ireland with no choice except to leave. For German immigrants, the beginning of the industrial revolution in the United States offered economic opportunity. Both groups arrived in New York at the dawn of the country's industrial surge and found a flourishing economy, with a demand for both unskilled and skilled labor. As hard as life in New York City proved to be, work was available, whether it was, as the Irish found, on the waterfront, or, for the Germans, as bricklayers and machinists in the city's first factories. The immigrants provided the city with an inexhaustible supply of labor for the port, and a willing workforce for the city's industrial revolution.

≈≈ The Industrialization of New York

The Port of New York did not serve primarily as a transfer point for arriving and departing cargo, as Liverpool did. England's industrial revolution began inland, in Lancashire, where textile mills filled the cities of Birmingham and Manchester and hundreds of smaller communities. A surplus rural population, driven off the land by the collapse of a cottage-based textile industry, provided the labor needed in the mills, not immigrants, as was the case in New York.

Often the pictures of America's new industrial might are the blast furnaces in the steel mills in Pittsburgh and, later, Henry Ford's assembly line. On the contrary, Manhattan was the leading manufacturing center in the United States, accounting for over 12 percent of the value of all manufactured goods by 1860, far more than any other city. Edwin Burroughs and Mike Wallace, in their award-winning book, *Gotham*, devote two large sections to the ascendency of New York as the center of the new American industrial economy: "Part Four: Emporium and Manufacturing (1844–1879)," and "Part Five: Industrial Center and Corporate Command Post (1880–1889)."[35]

On Manhattan Island, a combination of the city's Atlantic highway, the inland waterways, a large labor force, capital accrued in the maritime world, and aggressive entrepreneurs fueled the city's industrial growth. Relatively cheap transportation by water brought raw materials to the waterfront from around the world, across the country, and the nearby region. Manufactured products, in turn, could be distributed from the city to a vast hinterland. The arrival of the railroads to the port only reinforced the advantages that New York held as the center of a vast transportation nexus.

Lithographs of Manhattan Island from the 1880s illustrate a vibrant maritime world, with hundreds of ships at the piers and on the waterways in the port. No birds-eye view, however, can capture the growth of New York City's manufacturing activity, which paralleled its maritime economy. In 1880, manufacturing companies located on Manhattan Island employed 227,353 people, and by 1900, that number rose to 344,054.[36] The factories needed raw materials, and then finished products had to be distributed, not just in the metropolitan region, but throughout the country. In an age of steam power, the factories required coal, brought on barges to the closest pier, to fire their boilers. New York's manufacturing flowed across the waterfront, adding to the crush of freight.

The 1880 *Report on the Manufactures of the United States* ranked the leading cities in the country by industry, based on aggregate production. Manhattan ranked first, followed by Brooklyn.[37] Of the industries that were listed, New York ranked first for eleven and second for seven industries among the thirty-seven major manufacturing categories. The city ranked first in:

- blacksmithing
- bread and bakery products
- carpentering
- clothing (men's)
- clothing (women's)
- foundry and machine-shop work
- furniture
- malt liquor
- marble and stone work
- printing and publishing
- tobacco and cigars

It ranked second in:

- carriages and wagons
- lumber (planed)
- mixed textiles
- shipbuilding

- silk and silk goods
- slaughtering and meat packing
- tinware and copperware

The sheer diversity of the industrial companies located in Manhattan set New York apart from all of the other manufacturing cities.

New York's lead in manufacturing was driven by the technological revolution in the maritime world. The shift from sails to steam power provided a market for the foundries and machine shops along the East River, near the famous shipyards. At 13th Street, the Novelty Iron Works manufactured steam engines for the steamship revolution, as well as cast iron façades for the city's new commercial buildings.[38] The Novelty Iron Works' buildings stood just inland from the East River piers, where barges delivered coal and iron ore by way of the canals linking the port with eastern Pennsylvania. In 1851, the factory employed over 1,200 workers, and its furnaces ran day and night. Passengers on the Long Island Sound steamboats traveling on the East River often commented on the fiery beacons.

The Novelty Iron Works built engines for the steamboats on the Hudson River and Long Island Sound, as well as for some of the first transatlantic steamships, which were built in the East River shipyards. The early steam engines filled large spaces in the center of a boat: pistons, levers, and beams connected to the main shaft, which turned the paddlewheels. At the Novelty Iron Works, all of the separate components were molded, cast, finished, and assembled: a herculean task that took months. When launched at one of the nearby shipyards, a new steamboat would be towed the short distance up the East River to the Novelty Iron Works' dock, to have the engine installed. Due to the scale of its operation and the size of its workforce, Novelty Iron Works became "the best known industrial workplace in New York City and a symbol of industrial progress."[39] The East River shipyards and the Novelty Iron Works illustrate the synergy between technological change in shipping and the innovative manufacturing of steam engines.

New York City also ranked as the leading manufacturer of men's and women's clothing, and the garment industry would come to play a major role in the city's economy. The clothing industry connected directly with the cotton web. Many of the garments were made of cotton fabric that arrived in New York by ship: high-grade cloth, via Liverpool, from the English textile mills; and cheaper cloth from the Rhode Island and New England textile mills, which was carried on the Long Island Sound steamers' return trips. Cotton bales traveled from the South to the Manhattan waterfront and then to Liverpool and New England, to return as cloth to be made into garments marketed across the country. Clothing remained a major industry well into the twentieth century, when millions of Europeans arrived and found their first jobs in the sweatshops of the Lower East Side.

New York's ascendency in manufacturing only intensified the crowding along the shoreline in Lower Manhattan. By 1900, there were almost 40,000 manufacturing businesses on the island. Both the expansive Novelty Iron Works and a small sweatshop on Hester Street needed to obtain raw materials and transport finished goods, compounding the city's enduring transportation problems. Thousands of workers crowded the streetcars and ferries each working day, filling Manhattan's waterfront streets.

Manhattan's waterway empire drew the world's and the country's maritime commerce to the piers along the East and Hudson Rivers. Manufacturing only added to the pressure for space there. The piers along the East River, where the Port of New York began, could never accommodate the incessant demand for wharf space, so construction of new wharves and piers on the Hudson River followed. Even with the addition of made-land, which moved the shoreline out into the surrounding rivers, the demand for access to the Manhattan waterfront could never be adequately met, and this remained a critical problem well into the twentieth century.

[5]

The Social Construction of the Waterfront

The rise of the Port of New York depended on a large surplus labor force living along the shore. When men heard the cry "along shore men," they scrambled to the docks, hoping for one day's wages unloading or loading cargo. A day-labor system persisted on the waterfront, and a surplus of labor meant that the longshoremen had to compete with one another to be chosen for work at the shape-up. A ship's officer or, later, a dock boss or a criminal racketeer assembled the men at the entrance to the dock and then chose the number of men needed. The others drifted away to wait for another ship. With no guaranteed work or income, the waterfront drew the least-skilled laborers, often immigrants. A Russell Sage Foundation study of longshoremen, published in 1914, identified the shape-up and the uncertainty of steady employment as the source of the hardships the dockworkers confronted each day: "In the hiring we come at once face to face with the most conspicuous characteristic as well as the most far-reaching evil of the work— its irregularity. . . . Longshoremen have no way of knowing exactly when a ship will dock. Their only surety of being on hand for the hiring is to hang about, sometimes for hours until the ship arrives. . . . Nor is there any guarantee that the work will be permanent when obtained. . . . The stability of income necessary to maintain any settled standard of living whatever, does not exist."[1] The flood of immigrants coming to New York for over two centuries ensured a glut of labor in the port, available to be exploited. Desperate men who were willing to endure the uncertainty of the shape-up could always be found.

The longshoremen hired at the shape-up faced a hard day of dangerous labor. Climbing down into the holds, the men muscled the cargo into a sling, where it was lifted and deposited on the pier. More hands

Figure 5.1. Longshoremen, drays, and cargo made for a crowded pier on the East River. Sailing ships unloaded freight onto the piers, and horse-drawn carts (drays) hauled it off the piers. Longshoremen working the docks were picked in a shape-up. *Source*: *Harper's Weekly Magazine*, July 14, 1877

loaded wagons, called drays, that were waiting on the piers to move the freight to a warehouse, a factory, or another pier (fig. 5.1). The cargo destined for the railroads on the other side of the Hudson River was lifted from the holds and then placed onto flat-bottomed barges, or lighters, tied to the outboard side of the ship. A chaotic swarm of laborers filled the piers by day—and often into the night—and the harbor teemed with tugboats, lighters, and ferries moving the vast amounts of freight around the port. For ships departing from New York, the process was reversed. Drays delivered the outgoing freight, which was lifted aboard ship and then down into the hold, where the longshoremen stowed the cargo for the voyage.

The next day, longshoremen and pier workers moved on to find work on another ship. If nothing materialized, they drifted to the waterfront bars, to keep an ear open for news of a ship's arrival. The Russell Sage study did not overlook the sterotype of the longshoremen: addicted to drink and habituated to life in the waterfront saloons. But the study did point out that when men were not hired for the day, they needed to wait nearby. The bars on South and West Streets provided shelter, warmth, comradery, and, of course, alcohol: "At his best, the longshoreman is a fair sample of humanity—honest, generous, interesting, and individual. . . . At his worst, he is all his critics accuse him of being. But blame for the drinking, to which he must plead guilty, falls largely on the conditions of his work."[2]

Tens of thousands of workers made their livelihoods on the waterfront. This labor force included not only those men who worked the docks and drove the drays, but also the watermen who manned the tugboats, lighters, and ferries that traveled around the harbor. An army of men built and maintained the piers, dredged the slips, and worked in the hundreds of warehouses on the nearby streets.

No reliable data on the total employment generated by maritime commerce in New York's port during the eighteenth and nineteenth centuries exists. The Russell Sage study estimated that 35,000 men worked as longshoremen in 1914, and that number can be multiplied by a factor of ten to estimate the total maritime employment for that year.[3] Moreover, the estimates do not include the sailors who manned the thousands of ships—operating across the oceans and along the coasts and inland waterways—that had New York as their home port.

The waterfront neighborhoods were as essential to the growth of the port as its physical infrastructure: wharves, piers, and waterfront streets created by the water-lot grants. Longshoremen and dockworkers needed housing as close to the waterfront as possible while waiting for the next shape-up. Since they could never be sure of steady employment or income, they crowded South and West Streets each day. The tenements on the adjacent streets provided shelter for their families, no matter how wretched the conditions there were otherwise. A tribal way of life flourished, and the neighborhoods took on the aura of a rural village back in Ireland or, later, Sicily or wherever the home country of the next wave of immigrants had been.

～ Waterfront Workforce

Immigration fueled the rise of New York City to become America's entrepôt to the world, the leading industrial city in the country, a global center of finance, and the largest city in the United States. Immigration also provided the itinerant workforce needed for the port's maritime commerce. On Manhattan Island, the population grew from 33,131 in the first federal census in 1790 to 1,850,093 in 1900. New York remained the fastest-growing place in the United States, and, without immigration, the population could never have increased so rapidly. At the turn of the twentieth century, a number of the city's neighborhoods had some of the highest population densities on Earth.

The pathway for millions of immigrants coming to America led across the Atlantic Ocean—on sailing ships and, later, steamships—to the Manhattan waterfront. Thousands and then millions came to the United States for religious and political freedom and for economic opportunity, despite the difficulties of the voyage and the struggle to gain a foothold in America. By the middle of the nineteenth century, immigrants from Ireland and Germany composed over one-third of the city's population (table 5.1).[4]

The Irish, even from the beginning of their exodus to America and New York, found work on the waterfront. During the famine years, the most destitute settled in housing on the streets that were steps away from the piers along the East River, creating a distinctive Irish waterfront. The Russell Sage study reported that "as late as 1880, 95 percent of the longshoremen of New York . . . were Irish and Irish-Americans."[5] "Irish" referred to immigrants born in Ireland, and "Irish-Americans" to the first or second generation born in New York that had parents who came from Ireland.

Table 5.1.
New York City (Manhattan) foreign-born population from Ireland and Germany

Year	Population	Born in Ireland	Percentage	Born in Germany	Percentage
1845	371,233	96,581	26.0	24,416	6.6
1850	501,732	133,370	25.9	56,414	10.9
1860	801,088	203,740	25.0	119,964	15.9
1870	929,199	201,999	21.7	151,203	16.3
1880	1,222,237	200,300	16.5	163,482	13.4
1890	1,515,371	190,418	12.5	210,723	13.9

Sources: New York [State] Secretary's Office, *Census of the State of New-York, for 1845* (Albany: Carroll & Cook, 1846); for 1850–1890, US Bureau of the Census, *A Century of Population Growth from the First Census of the United States to the Twelfth, 1790–1900* (Washington, DC: US Government Printing Office, 1909).

By comparison, few German immigrants worked the waterfront. The Germans who came at the same time as the Irish were not impoverished, had much higher levels of education, and often worked at a skilled trade.[6] The Germans found steady employment and did not have to face the shape-up to compete with others each and every day, as the Irish did. In addition, a number of German immigrants brought savings with them, which they used to open small retail, construction, and manufacturing businesses. The Germans who settled in the tenements in the Lower East Side in the 1850s stayed for a comparatively brief time period. As soon as they could afford to, they left the Lower East Side for Yorkville, on the Upper East Side of Manhattan, moving Kleindeutschland (Little Germany) far from the slums along the East River.

A spatial analysis of the 1880 census data,[7] the first federal census to include household addresses, illustrates the ethnic makeup of four waterfront neighborhoods settled by the Irish: two along the East River (Wards 4 and 7) and two on the Hudson River (Greenwich Village and Chelsea). German immigrants clustered in Wards 11 and 17 on the Lower East Side, below East 14th Street, around Tompkins Square, which formed the center of Little Germany (map 5.1).

The 1880 census data for the waterfront neighborhoods and Little Germany illustrates the process of chain migration: newly arriving immigrants moved to neighborhoods where their countrymen already lived. Formal and informal systems of communication encouraged the newly arrived to find lodging where they would be welcomed. Clustering near relatives who had already settled on a block created dense ethnic neighborhoods.[8] Hundreds of thousands of immigrants arrived each year, filling the worst of the city's housing. The poorest of the poor had no choice but to live in atrocious conditions, and landlords had no incentive whatsoever to make improvements. If one family moved out, another trudged up the streets from the piers and moved in.

The Irish, both those born in Ireland and their first-generation children born in the United States, made up over a third (34.2%) of Manhattan's population.[9] In Wards 4 and 7, on the East River, the Irish totaled over 50 percent of the residents. In Ward 17, on the Lower East Side, the ethnic pattern was reversed: the Germans formed the majority of the population, with few Irish residents. On a smaller spatial scale, on the blocks adjacent to Tompkins Square in Ward 17, the ethnic segregation becomes even more striking. Manhattan, with only 18 per-

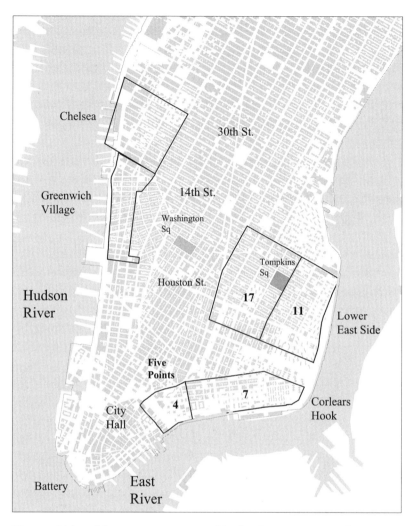

Map 5.1. Irish and German immigrant neighborhoods in 1880. The Irish were in wards 4 and 7 (on the East River), as well as in West Greenwich Village and Chelsea (on the Hudson River). The Germans were in Kleindeutschland, or Little Germany (wards 11 and 17). *Source*: Created by Kurt Schlichting. Source GIS layer: building footprint map, New York City Planning Department.

cent of its population being native born, constituted a world apart from the rest of America, where the native-born made up 86.7 percent of the US population of 50 million in 1880. New York, during the century of immigration, became a populous multiethnic enclave, a process that continued up to the turn of the twentieth century, when ethnic tensions arose and nativism triumphed, leading to drastic immigration restrictions in 1924.

≋ The Irish Waterfront

Burroughs and Wallace state that by the middle of the nineteenth century, the Irish "took over the docks" along the East River. Each day, between 5,000 and 6,000 longshoremen filled the piers along the East River for the shape-up: "The work was hard, poorly paid and erratic. While waiting for ships to arrive or weather to clear, men hung around the local saloons, took alternative jobs as teamsters, boatmen or brick makers, and relied on the earnings of their wives and children."[10] In the Irish waterfront neighborhoods, economic stability remained elusive.

East River, Ward 4

Many of the famine Irish settled along the waterfront on the East River, in Ward 4. After a voyage lasting two months or longer, the immigrants simply walked off the ship onto the piers lining South Street and carried their meager belongings a block or two up Roosevelt, James, or Oliver Streets to a tenement or a cheap boarding house.

The area along the East River around Roosevelt Street, Cherry Street, Pecks Slip, and Franklin Square formed part of Ward 4, where the 1850 US census reported a total population of 23,123.[11] Chatham Square formed the northern boundary and Ward 4 included all streets down to the waterfront. An 1881 map of the area shows the construction of the approaches to the Brooklyn Bridge on Franklin Square (map 5.2). From there, Cherry Street runs east and crosses Roosevelt, James, and Oliver Streets. Men from Ward 4 worked the docks and at the Fulton Fish Market, which supplied the city with fresh seafood.

The 1880 census subdivided Ward 4 into eleven enumeration districts (EDs), and the area along Cherry Street formed ED 33. In that ED, the population totaled 1,788 residents, but only fifty-one individuals (2.9%) were native, meaning they and both of their parents were born in the United States. By comparison, over 74 percent of the population in the ED was Irish, either born in Ireland or first-generation (born in the United States). In Manhattan, the German immigrants nearly equaled the Irish in numbers, but only eighty-nine Germans lived in ED 33. Intense ethnic segregation characterized Ward 4 and other Irish

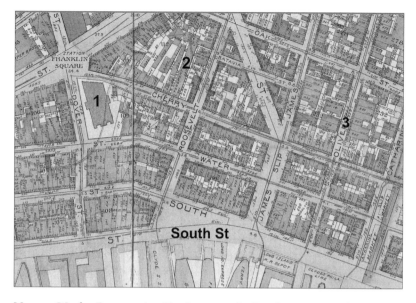

Map 5.2. Ward 4, Enumeration District 44, on the East River. The numbers on the map indicate the foundation for the Brooklyn Bridge (1); Gotham Court (2), at 34 and 36–38 Cherry Street; and Oliver Street (3), the childhood home of Al Smith, a governor of New York and Democratic candidate for president of the United States in 1928. *Source*: Created by Kurt Schlichting. Source GIS layer: historic street map, New York Public Library, Map Warper.

waterfront neighborhoods, as it did in Kleindeutschland, farther to the north in the Lower East Side.

Alfred Smith, the famous Tammany politician, governor of New York, and the first Catholic to run for president of the United States in 1928, was born in 1871 on Oliver Street, in Ward 4. His mother's parents immigrated to New York from Ireland, and his father's parents came from Italy and Germany. Smith described his childhood as growing up in the "shadow of the Brooklyn Bridge."[12] An early biography of Al Smith described Ward 4 in lurid detail: "This book is the story of a boy who grew up in a region where Tammany Hall was supreme. The Fourth Ward and the parish of St. James in which he lived were surrounded by poverty and vice in terrible forms. . . . He is the first of our national heroes to be born amidst din and squalor. . . . Life on the East Side of New York when Alfred Smith was born was an even fiercer struggle for existence than it is today [1927]."[13] While successive waves of immigrants would come to Ward 4 from across Europe, "before any of these large tides of immigration, there arrived the Irish."[14]

Residential buildings in the neighborhood, originally constructed for a single family, were quickly subdivided into tiny apartments, like the one Al Smith's family lived in on Oliver Street. Landlords often built a second shoddy structure in the rear yard, to cram in more immigrants. The wretched housing available to the Irish included some of the most shameful in the city and led to reform efforts involving well-to-do New Yorkers appalled by the Tammany corruption and the complete failure of the city government to respond. In the 1850s, Peter Cooper and a group of wealthy activists formed the Citizens' Association of New York. The association's first major initiative focused on public health and demanded that the New York Legislature establish a health department independent of Tammany Hall. In Albany, the political allies of the association in the legislature introduced a bill in 1859 to form a health department, but that failed, given the fierce opposition from the Tammany machine. The Civil War intervened, but the draft riots in 1863 strengthened the reformers' conviction of the need for "social hygiene" in the immigrant tenement districts, such as Ward 4. The alternative promised continued chaos and social unrest. Finally, in 1866, the legislature passed a bill establishing the independent Metropolitan Board of Health for New York City and the adjacent counties. The law required that at least three of the commissioners were to be physicians, not Tammany political hacks.

While waiting for the state legislature in Albany to act, in 1864 the Citizens' Association funded a major study of the social and health conditions in the city's neighborhoods, including Ward 4. The association's Council of Hygiene and Public Health completed a study of the "sanitary condition of the city."[15] The Council divided the city into twenty-nine sanitary districts. In each district a research team, headed by a doctor, conducted an exhaustive door-to-door survey of the tenements and the people living in them. The investigators gathered health data and inspected 15,511 tenements, which were home to 486,000 residents, over half of the city's total population of 813,669 in 1860. In Ward 4, the study identified 486 tenements, with 17,957 residents, or three-quarters of the ward's total population of 21,944 enumerated in the 1860 US census.

Ezra E. Pulling, MD, authored the report for the Fourth Sanitary District, which included the entirety of Ward 4. This sanitary report documented in grim detail the appalling conditions in the immigrant neighborhoods. Pulling listed the streets in the district that had no sewers and then added that those streets with sewers "empty into the East River." He noted that most of the human waste in a city of over 800,000 emptied into the East or Hudson Rivers, and, if not, "liquid refuse [human waste] is emptied on the sidewalk or into the street, in some instances into sinks [underground vaults] in the domiciles communicating with a common pipe which discharges its contents into an open gutter to run perhaps a hundred feet, giving forth the most noisome exhalation, and uniting its fetid streams with numerous others from similar sources."[16]

In an era before the acceptance of the germ theory of disease, Pulling and the other physicians believed the cause of disease to be miasma, or "bad air." Specifically, this fetid air emanated from privies, open latrines, and sewers and was combined with the lack of ventilation in the crowded tenements. The reports for the other twenty-eight sanitary districts all focused on bad air in the tenements, privies, and the city's rudimentary sewer system. All of the reports claimed that this miasma caused the appalling health of the city's poorest residents, who were forced to live in the tenements.

Of the 1,507 buildings in Ward 4, 682 were businesses, 8 were churches or schools, 47 were stables, and 770 were primarily residential buildings. The investigators not only counted but also categorized the residential buildings, classifying 740 of them as "tenement[s] and crowded houses." Among these, 472 were single-family homes that had been converted to tenements, and 242 were built as tenements. Of the 770 residential buildings, 96.1 percent were identified as overcrowded tenements, and 108 of them included a second tenement building in the rear lot.[17]

The Pulling report describes a typical tenement house, using a dwelling on Jane Street as an example: a brick building, five stories tall, with a hall 4.5 feet wide; stairs 2.5 feet wide, in perpetual darkness; and four two-room apartments on each floor. One room measured 14 × 12 feet and the second 9 × 12 feet, with a coal stove and no window. The bedroom contained two beds and bedding on the floor. It was "a dormitory of half a dozen persons. A sickening and stifling odor, most offensive to the unaccustomed sense pervades this apartment and poisons the atmosphere inhaled by the residents."[18]

Gotham Court

Charles Dickens visited New York in 1842 and toured Five Points, the infamous Irish slum north of Ward 4, where Chinatown is today. He continued his journey south into Ward 4 and visited one of the most notorious tenements in Manhattan: Gotham Court, on Cherry Street (map 5.2, #2). For almost fifty years, thousands of Irish immigrants and their children resided in Gotham Court. The men living there found work on the piers lining the East River.

In 1850, Silas Wood, a wealthy landowner in Ward 4, constructed a massive tenement complex, known as Gotham Court, on Cherry Street, between Roosevelt Street and Franklin Square. Wood built two buildings—numbers 34 and 36–38 Cherry Street (fig. 5.2). Each stood six stories high, 250 feet in length by approximately 35 feet wide. Apartments on each floor measured 10 × 14 feet, divided into two rooms.[19] Two alleys, East and West Gotham Place, ran along the side of the build-

Figure 5.2. Gotham Court, at 34 and 36–38 Cherry Street, was three blocks from the East River. This crowded, stinking slum was one of the many tenements that were home to the Irish longshoremen who worked the docks. *Source*: Museum of the City of New York

ings, allowing access to the stairwells. Privies and sinks were located in the cellars, which required the residents to climb up and down stairs. Grates in the alleys provided some ventilation for the reeking privies.

Pulling's report found seventy-one families living in Gotham Court, with a total of 504 residents, an average of seven persons for each cramped two-room apartment. The filthy apartments stank, due to "no small portions of their own excretions, as it is painfully evident to everyone who has . . . an acute sense of smell."[20] Among the families who lived in the Court in the previous two years, there had been 138 births, including twelve babies that were stillborn. Only seventy-seven infants survived to the age of 2, a mortality rate of over 44 percent. To a modern sensibility, the number of infant deaths in Gotham Court would be a health crisis. For the mothers and fathers who watched their babies die, their grief must have been unbounded. Only the poorest Irish immigrants lived in the Court, as evidenced by the forty-nine vacant apartments. They must have recognized the danger to their children's health, and those with means left, even if they moved to another tenement a block away. Pulling's report also included detailed descriptions of horrific conditions in other Ward 4 tenements. He argued that while Gotham Court had a reputation as one of the worst, "on the whole . . . Gotham Court presents about the average specimen of tenant-housing in the lower part of the city. . . . There are some which are more roomy, have better ventilation and are kept cleaner; but there are many which are in far worse condition."[21]

Obtaining decent food proved to be especially challenging. The small grocery stores, including one in Gotham Court that faced Cherry Street, sold ghastly food, which was often tainted and "unfit for human consumption." In addition to the spoiled food, the stores sold the then-legal narcotics that many used to dull their senses. The grocers, who often also owned tenements in the neighborhood, mercilessly overcharged their customers. With so little income, the residents often bought on credit and found themselves in debt for both rent and food, held in "a state of abject dependence and vassalage little short of absolute slavery."[22]

In Ward 4, 134 small groceries sold food, with one shop for every twenty-seven families. The number of "drinking places" totaled 446, with one for every eight families. In addition to numerous saloons on every block, there were twenty-eight brothels and six "sailor's dance-houses." Herman Melville's autobiographical novel *Redburn* includes vivid descriptions of the bars and brothels along the East River waterfront in Ward 4.

Pulling ends his sanitary report with a surge of moral outrage at the horrid conditions his investigations uncovered. He foreshadows Jacob Riis and other reformers who demanded that the city's government respond to save the lives of people condemned to live in Gotham Court, Ward 4, and the other immigrant neighborhoods. Pulling called for the removal of one-half of all the tenements and demanded that if private interests would not build decent housing, it "should become a subject of municipal and legislative action."[23] It would be half a century before a government, at any level, even imagined an obligation to ensure decent housing for the poorest of citizens and, in the case of New York City, the millions of immigrants who would follow the Irish and Germans.[24]

An 1879 article about Gotham Court in *Harper's Weekly Magazine* included direct quotes in the native brogue of the Irish residents describing the difficulties they faced. One resident, who paid $4 a month in rent, lamented about her desperate situation:

> Nearly every room on the lower floor was found to be like those just described—small, dark, and dirty—and every tenant had her grievances which she poured forth at the first opportunity. *"Bad luck t' them,"* said one. *"I fill intil the cellar lasht week—how cud I help it, Sir—and hurt meself bad, so I did."* She was a sad-faced creature. . . . She was obliged to live there, she said, because she was desperately poor. *"Not a cint comin' in but tin cints* [ten cents] *a noight from two girls as lodges wi me, and whatever 'll I do now? Wan o' them's jisht come in from work wi her fut shprained. She's lyin' on the bed in there."*[25]

The article in *Harper's* depicts Danny Burke, at 4 years old the youngest newsboy in New York City, who sells the *Evening Telegram* paper on the corner for 2 cents. The writer meets Danny in a stairwell and asks a woman, who praised the boy's hard work, if Danny was her son. *"No, Sir,"* she answered, *"he belongs to the woman up shtairs, but I've wan insoide that 'll match him. Luk at that, Sir,"* and she pointed to a little one sleeping in the cradle. A man standing in the hallway told the reporter that the woman *"had twelve o' them, Sir."* The reporter asked the woman,

"Are they all living?" She thought for a while and then related sadly, "*No, Sir, risen* [seven] *o' than are dead.*" Seven of her twelve children had died in infancy while living in Gotham Court.

The 1880 census found seventy-two families living in Gotham Court—almost the same number as in Pulling's sanitary report a decade earlier—but counted far fewer residents: 299 versus 504. The average family size in 1880 was 5.5, versus 7 a decade earlier.[26] Yet even with fewer residents and smaller families, Gotham Court still remained crowded, where a family of four or five shared two small rooms. Among these seventy-two families, fifteen women headed households, all of them widows. Elizabeth Daly, age 42, and Mary Mulcahey, age 52, both Irish immigrants, lived next door to each other. Elizabeth's three children—Ellen, age 29; Mary, age 26; and Peter, age 21—all worked to provide the family with income. Both girls, who were unmarried, worked in a paper-box factory, and Peter found low-paid employment as a day laborer. Mary Mulcahey had a 10-year-old daughter who attended school, most likely St. James Catholic School around the corner, Al Smith's grammar school. No employment was listed for Mary.

Three blocks down the hill from Cherry Street, the piers on South Street teemed with maritime commerce. Longshoremen from the neighborhood gathered there early each morning, in all seasons, for the shape-up. Among the fifty-seven men with families living in Gotham Court, fourteen worked as longshoremen, almost 25 percent. All of the longshoremen except John Sullivan, born in New York in 1830, were Irish immigrants. Both of Sullivan's parents had emigrated from Ireland before the potato famine. The Irish longshoremen who had families and were living in Gotham Court included James Casey, age 38, with four children: ages 6, 4, 2, and a new baby born in 1880. The census listed another longshoreman resident, James O'Connell, age 54, as not able to read or write. He and his wife Margaret had five children. The three oldest—Johanna, age 23; Patrick, age 20; and Daniel, age 17—all worked. Margaret remained home with their daughter Margaret, age 13, and son John, age 11. Only John attended school. The longshoremen's wives remained at home, but in all the families, the older children had to work down on the docks to provide additional income to pay expenses, including the rent at Gotham Court.

Given the dangerous nature of the longshoremen's work, injuries were common. Daniel Murphy, who broke his jaw working the piers, reported no income for five months. His neighbor, John Costello, was out of work for four months. Daniel and his wife Mary had one child—Mary, age 20—who worked in a paper-box factory nearby and provided the only reported income the family had for part of the year. Few residents of Gotham Court had good jobs that paid a decent income. If they did, they would have left Cherry Street for more-decent housing.

Jacob Riis, in *The Battle with the Slum*, refers to Gotham Court as "the tenement that has challenged public attention more than any other in the whole city and tested the power of sanitary law and rule for forty years."[27] The owners refused to make any improvements to protect the health of the residents. The health inspector for the district reported that "nearly ten percent of the population [of Gotham Court] is sent to the public hospitals each year."[28] While advocating for change, Riis, in his powerful *How the Other Half Lives*, also offers the worst stereotypes to describe the poor Irish residents there: "A Chinaman whom I questioned as he hurried past the iron gate of the alley, put the matter in a different light, 'Lem Ilish velly bad,' referring to the Irish immigrants and their children who lived in the Court."[29] Despite the damning sanitary reports, the *Harper's* story in 1879, and Riis's *How the Other Half Lives*, Gotham Court would not be demolished until 1895. The Irish continued to fill the apartments in Ward 4 for decades, including the notorious slum on Cherry Street.

Writing in 1902, Jacob Riis reminds readers of just how horrific life had been in this rookery:

> Look at Gotham Court, described in the health reports of the sixties as a "packing-box tenement" of the hopeless back-to-back type, which meant that there was no ventilation and could be none. The stenches from the "horribly foul cellars" with their "infernal system of sewerage" must needs poison the tenants all the way up to the fifth story. I knew the Court well, knew the gang that made its headquarters with the rats in the cellar, terrorizing the helpless tenants; knew the well-worn rut of the dead-wagon and the ambulance to the gate, for the tenants died there like flies in all seasons, and a tenth of its population was always in the hospital.[30]

Riis condemned the moral failure of the landlords who owned the horrific tenements and who fought every reform effort: "Think of the living babies in such hell-holes; and make a note of it, you in the young cities who can still head off the slum where we have to wrestle with it for our sins. Put a brand upon the murderer who would smother babies in the dark holes and bedrooms. He is nothing else."[31] Emphasizing the dignity of every human being, no matter how poor, in moral and religious language, Riis's words would find an echo in calls for change on the waterfront in the twentieth century. The longshoremen who still lived in the dank tenements deserved the right to better working conditions on the piers, where the shape-up, the loan sharks, and the mobsters made a mockery of man's basic human right to dignity.

The Hudson River Waterfront

The city's maritime world had its beginning along the East River, but the unceasing demand for pier space stimulated the development of the Hudson River waterfront. The sizes of ships coming from Liverpool, Europe, and elsewhere around the world had increased dramatically. Much longer piers could be built along the Hudson, and transatlantic shipping, for both cargoes and passengers, moved to the West Side of Manhattan. When the Department of Docks took over, the original piers on the Hudson River were already obsolete. The department removed the old piers and constructed much longer piers, in order to accommodate the new generation of ships. The development of the new piers, warehouses, and factories on the adjacent blocks drew more shipping—and the railroads—to the Hudson River side of the island. The railroads and the transatlantic steamship companies signed long-term leases with the Department of Docks, to ensure access to the waterfront.

The history of Greenwich Village and Chelsea goes back to colonial times, when the area offered a bucolic retreat for wealthy residents seeking to escape the crowded, unsanitary conditions in the city, which then was a mile to the south at the tip of Manhattan Island. With the shift in shipping to the Hudson River, the Irish longshoremen followed, and both Greenwich Village and sections of Chelsea became Irish enclaves (see map 5.1). James Fisher's account of the Irish waterfront chronicles the transfer of port activity first to the Greenwich Village waterfront and then north to Chelsea: "The focus [of the Department of Dock's rebuilding of the piers] soon shifted almost entirely to the Lower West Side. . . . In 1897 the department supervised construction of municipally owned piers—upwards of seven hundred feet long—on riverfront terrain between Charles and Gansevoort Streets. The West Side's Irish waterfront coalesced around the imposing new structures before expanding northward up the Hudson shoreline [to Chelsea]."[32] Just as in Ward 4, residential buildings on the streets leading down to the Hudson waterfront, originally built for one family, became tenements. Both the Village and Chelsea had more native-born residents than the city as a whole, and more ethnic diversity than either of the Irish neighborhoods along the East River.[33]

Greenwich Village

The Greenwich Village waterfront included the blocks west of Hudson Street, extending to the river between Houston and West 14th Streets (map 5.3). In response to the number of Irish Catholics moving to Greenwich Village, the Archdiocese of New York established St. Veronica's parish in 1887. An old stable on Washington Street served as the church during construction. The first mass was celebrated in 1903 in the new church at 153 Christopher Street (map 5.3, #3). St. Veronica's played an important role on the Irish waterfront.

In the Village, the neighborhood became more Irish on the blocks closest to West Street and the waterfront. The first blocks inland from the Hudson River—West Street to Washington or Greenwich Streets— had a much higher concentration of Irish than blocks farther east (map 5.4). In addition, ED 218, between Washington and Greenwich Streets, just south of Leroy Street, had the highest percentage of Irish in the Village, 60.8 percent. Living close to the Hudson provided the Irish longshoremen and dockworkers with quick access to the Christopher Street piers. Stuart Waldman's study of the Greenwich Village waterfront confirms that as soon as the Department of Docks completed the new Hudson River piers, hundreds of new tenements were built on the adjoining streets, to which the Irish moved.[34]

The 1865 report for Sanitary District 5B included the blocks from Houston to Christopher Streets along the Hudson River. Some of the tenements, once single-family homes, "are now crowded [with] from four to six families, averaging five persons each."[35] North of Christopher Street, Sanitary District 11 included the blocks up to West 14th Street. The streets near Gansevoort were described as "always filthy." The deficiencies of the sewerage system, especially the sewer under Gansevoort Street, which did not empty into the Hudson River, allowed hu-

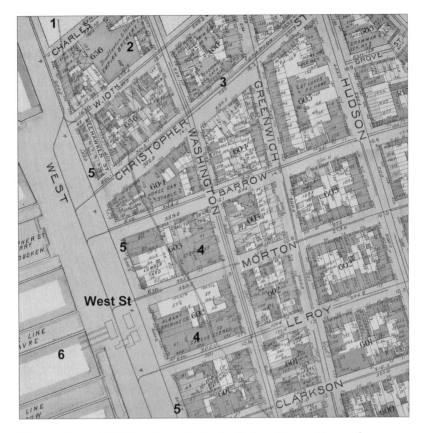

Map 5.3. Greenwich Village, from Charles to Clarkson Streets. The numbers on the map indicate Charles Place (1), known as "Pig Alley"; the Empire Brewery (2); St. Veronica's Catholic Church (3); warehouses (4); waterfront bars (5); and Hudson River piers (6). *Source:* Created by Kurt Schlichting. Source GIS layer: historic street map, New York Public Library, Map Warper.

man waste to flow into an open trench. The residents living on the blocks around the open trench—Gansevoort, West, Washington, and Horatio Streets—suffered "no less than 29 cases of dysentery and diarrhea, 5 of which had terminated fatally" in one three-week period.[36]

Once established in Greenwich Village, the Irish dominated its waterfront for the next forty years, even as immigration patterns to the city changed dramatically. In the latter half of the nineteenth century, the number of immigrants from Ireland and Germany declined, while the numbers arriving from eastern and southern Europe increased.

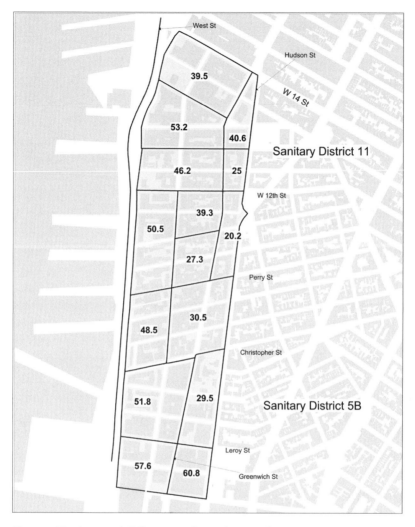

Map 5.4. The Greenwich Village waterfront, showing the 1880 census enumeration districts. The numbers indicate the percentage of Irish in each outlined area, which includes those born in Ireland plus first-generation Irish (born in the United States, but whose parents were born in Ireland). *Source*: Created by Kurt Schlichting. Source GIS layer: building footprint map, New York City Planning Department. Other sources: 1880 enumeration district map, Milstein Division, New York Public Library; "NAPP 1880 US Census, 100% Enumeration" [with individual-level data aggregated to the 1880 enumeration districts], Steven Ruggles, Katie Genadek, Ronald Goeken, Josiah Grover, and Matthew Sobek, *Integrated Public Use Microdata Series: Version 6.0* [database] (Minneapolis: University of Minnesota, 2015), North Atlantic Population Project, https://www.nappdata.org/napp/.

The Germans in the Lower East Side moved out to Yorkville, to be replaced by millions of eastern European Jews fleeing persecution and poverty in the Austro-Hungarian and Russian empires. In southern Italy and Sicily, life became increasing precarious, with few economic opportunities. America beckoned as the Promised Land. Young men from Italy traveled alone in the 1890s to find work in New York, and then either brought their families over or returned home. Despite the influx of these new immigrants, the Irish held on to their waterfront jobs and remained in the Village waterfront.

Over time, as the ethnic makeup of Manhattan changed, Greenwich Village did not. Between 1880 and 1920, over 35 percent of the population was born in Ireland or had parents who were.[37] For Manhattan as a whole, the percentage of Irish declined, from over 30 percent of the population in 1880 to 12 percent by 1920. Yet the tribal world on the Irish waterfront persisted.

Chelsea

Chelsea is generally considered to be the area that runs north of West 14th Street to West 26th Street, although the neighborhood's boundaries have always been a matter of debate. Clement C. Moore's property along the Hudson River above Greenwich Village long remained an undeveloped part of Manhattan (see map 2.4). With the dramatic increase in the size of ships coming to the port, when the Department of Docks took over the waterfront in 1870, they realized that the Chelsea section provided the most practical location for the massive piers needed to accommodate the large White Star Line and Cunard Line ocean liners. Their foresight was confirmed when, decades later, the opening of the Chelsea piers in 1902 signified a new era in transatlantic travel.

Even long before these piers were constructed, increased maritime activity on the waterfront drew the Irish to the Chelsea neighborhood. They also found work in the iron and brass factories, such as the Williams Ornamental Brass and Iron Works on West 26th Street and the Colwell Iron Works, around the corner on 11th Avenue, between 27th and 28th Streets, which were part of the city's thriving industrial economy.[38] The Consolidated Gas Company, which employed many immigrants, occupied four blocks: from 16th to 20th Streets, between 10th and 11th Avenues. An 1891 Bromley map depicts the early development of the shorefront before the construction of the Chelsea piers, when 13th Avenue would disappear, to be replaced by West Street (see map 7.3).

In the rush to construct tenements in Chelsea as the waterfront expanded and factories and train yards filled the adjoining blocks, landlords and developers crammed as much housing as possible on each block. The 1865 sanitary report labeled this the "massing of tenements." On each lot, a tenement occupied the street front, and a second building filled most of the rear of the lot, with the privies in between. A typical city lot in Chelsea measured 30 × 100 feet. On a single block, developers built twenty houses, "each twenty feet wide, and as high as it pleases the owner to rear them . . . on a space less than 20,000 square feet; and allowing each front house to contain eight, and each rear house four families (a modest estimate), we have each family [living in] about 164 square feet of ground."[39] East of 9th Avenue, on 26th Street, where the Irish lived, the report included a drawing of the massed tenements, labeled "Bird's-Eye View of a New Fever-Nest on the Avenues."[40]

The social geography of Chelsea mirrors that of Greenwich Village, with the Irish concentrated on specific blocks: 9th and 10th Avenues between 14th and 17th Streets, and along West Street and the Hudson River from 25th to 29th Streets. In these EDs, the Irish constituted over 70 percent of the residents. At the time of the 1865 sanitary report, rank tenements filled parts of Chelsea, particularity along the Hudson around West 26th Street, in ED 341. The survey counted a total of 1,154 residential buildings and tenements. In the more affluent sections of Chelsea, where the Irish did not live, new residential buildings included "all the modern improvements, such as the introduction of [manufactured] gas, Croton-water, baths, furnaces, water-closets and proper sewer connections."[41] "Croton-water" referred to New York City's new, massive public water system that drew fresh water from a series of reservoirs in Westchester County, north of the city. Water-closets included a flush toilet. By comparison, the sanitary conditions in the tenements in the Irish sections, even those that were recently constructed, remained "defective," with no connections to the sewers. In one building on West 22nd Street, forty-two people shared privies in the damp and dark cellar, with no ventilation, which was described as "worse than a Stygian pit."[42]

In a developing neighborhood like Chelsea, the social makeup could change significantly from block to block, and some of the streets included luxurious single-family homes for the wealthy and the family's three or four live-in servants. Clement Moore leased his land from 9th to 10th Avenues, between 22nd and 23rd Streets, to William Torey, who built eighty four-story brownstone townhouses in 1845, each leased as a private residence. The General Theological Seminary occupied the block two streets to the south. Before the Great Depression a developer, Henry Mandel, took over the leases from the Moore descendants, tore down the townhouses, and constructed the high-rise London Terrace building, with over 1,700 apartments, which attracted well-to-do business executives and their families.[43]

Despite upscale developments like London Terrace, sections of the Chelsea waterfront remained an Irish enclave into the post–World War II era. The International Longshoreman's Association, the notorious union that represented the waterfront laborers, located its first headquarters on West 14th Street. In the years to come, the pastor of Guardian Angel Church on 10th Avenue and the Jesuit pro-labor priests who taught at Xavier High School on West 16th Street would wage an epic battle for what Fisher called "the soul" of the Irish waterfront.[44]

[6]

The Port Prospers, the Railroads Arrive, and Congestion Ensues

The success of the packet lines and the expansion of New York's waterway empire solidified the Port of New York as the most important in the country. As the city's population continued to grow and its industrial bases expanded, demand for space on the waterfront increased exponentially. An ever-increasing number of ships arrived and departed each year, and thousands of canal barges brought the bounty of the Midwest to the port. Hundreds of small sailing vessels carried food and building materials from the nearby hinterland. For example, the US Bureau of Customs kept detailed records of vessels entering the United States directly from foreign countries. Between 1845 and 1860, the number of such ships rose from about 2,000 a year to over 3,500 in 1850, and then to 3,962 in 1860.[1]

Albion compiles data comparing foreign and coastal arrivals in 1835 to New York and to its closest rivals, Boston and Philadelphia. A total of 240 "ships," at that time meaning three-masted sailing vessels, arrived from Liverpool: 168 to New York, 38 to Boston, and 34 to Philadelphia, making New York's share 70 percent. From London, 50 ships arrived, with 44 going to New York. For the famed China trade, where Boston and Salem once dominated, 32 ships arrived on the East Coast from Canton, and 24 of these sailed through the Narrows to piers on South Street. New York also dominated the coastal trade routes to the American South, including the cotton ports of Charleston, Savannah, Mobile, and New Orleans. Ships arriving from the four southern ports totaled 1,066, with 712 coming to the Port of New York.[2]

In 1852, the *New York Times* began to publish the "Marine Intelligence" page almost daily, detailing arrivals and departures. The list, organized by type of vessel, included information identifying either the port the ships arrived from or their destination. New York's merchants and shippers read the report with great care, to stay abreast of their competition. The "Marine Intelligence" page illustrates the global and coastal reach of New York's maritime commerce by the 1850s. For example, on June 23, 1852, two steamships departed for Charleston and Glasgow, Scotland. Six sailing ships (three-masted) cleared for Spain, London, Charleston, Northern Ireland, and St. Stephen in New Brunswick, Canada.[3] Thirteen smaller barks and brigs set sail for Cuba; Gibraltar; Mexico; Bermuda; Bremen, Germany; Bergen, Norway; Kingston, Jamaica; Nova Scotia; and Philadelphia. Thirteen schooners left for ports up and down the coast from Boston to Jacksonville, as well as one to Harbour Island in the Bahamas and another to Port-au-Prince, Haiti. In addition, ships docked at the port from other distant shores: Bremen, Marseilles, Rio de Janeiro, and Puerto Rico. Coastal ships included the *Kelly* from Savannah (a voyage of seven days) and the *Moses* from Charleston (five days in transit), both loaded with cotton either to be transferred to the packet ships bound for Liverpool or shipped via Long Island Sound to the New England cotton mills. Along with the largest of the sailing ships, thirty-one smaller schooners also arrived.[4]

The *Alliance* from Le Havre, France, with 348 passengers, made port after a trip of thirty-six days across the Atlantic Ocean. The *Milicot*, a British ship from Dublin, carried salt and 299 passengers. The Black Ball ship *Isaac Wright*, thirty-five days out of Liverpool, brought another 531 passengers. On this one June day in 1852, 1,178 people landed at the East River piers, almost all of them immigrants. Although they arrived on a crowded waterfront, filled with goods from all over the United States and the world, these passengers may have stopped and realized they had passed through the "golden door," Emma Lazarus's evocative description of New York, inscribed on the Statue of Liberty.

With thousands of ships in the port on a daily and weekly basis, the demand for pier space far exceeded the amount available. A New York State study of the harbor reported that on almost any given day, 300 ships waited for space on the waterfront. An 1879 lithograph of the port and the Manhattan waterfront illustrates the scale of its maritime activity (fig. 6.1). In the foreground are sailing ships and a steamer

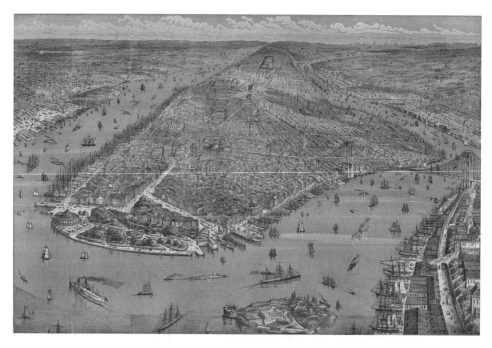

Figure 6.1. The Port of New York and the Manhattan waterfront, 1879. By the 1870s, the port was the second busiest in the world. Hundreds of ships of all types and sizes crowded the harbor, the East and Hudson Rivers, and the Brooklyn waterfront (*in the foreground*). *Source*: Rogers, Peet & Company, 1879, Library of Congress

heading down the East River from Long Island Sound. Three-masted sailing ships occupy many of the piers along the Hudson and East Rivers, and hundreds of small sailing vessels ply the waters around the island. Every pier along the East and Hudson Rivers is filled. The Brooklyn Bridge connects Manhattan to the city of Brooklyn, and the piers along Brooklyn Heights to the south of the bridge are full.

～ The Railroads Cometh

Not only did the world's shipping make its way to New York, but the railroad era also drew all the country's major rail lines to the port. After his shipping triumphs, Cornelius Vanderbilt dominated New York State's railroad history in the second half of the century.[5] Vanderbilt did not build railroads, he acquired them. His two most important acquisitions were the New York & Harlem Railroad and the Hudson River Railroad. Both of these had an incredibly valuable asset: their charters gave them the right to operate on Manhattan Island and to build their tracks in the middle of the city's streets.

In 1831, the State of New York awarded a franchise to the New York & Harlem to build a railroad to connect the southern tip of Manhattan Island with the village of Harlem, today the area around 125th Street.[6] The New York & Harlem's first trains used horses for power, and its tracks ran up 4th Avenue, later renamed Park Avenue. In 1837, the railroad finally reached Harlem. The state legislature then extended its franchise to White Plains in Westchester County and, later, to Chatham in Columbia County, to join the Boston & Albany Railroad. Although never a prosperous railroad, the Harlem had one superlative asset: the right to operate on the island into Midtown Manhattan.

In 1846, the state legislature granted a charter to a group of investors from Poughkeepsie, New York, to build a railroad down the east side of the Hudson River to New York City. The Hudson River Railroad crossed the Harlem River at the northern tip of the island and then went all the way down the West Side to St. John's Park Depot, blocks from city hall (map 6.1). Like the New York & Harlem, the Hudson River Railroad held the right to use the city's streets along the river, an asset as valuable as the New York & Harlem's Park Avenue rights. The Hudson River Railroad built tracks down the middle of 11th Avenue, 10th Avenue, and then south on Hudson Street to St. John's Park Depot. The city had previously granted water-lots to private individuals to develop the port's maritime infrastructure. Then, as the railroads arrived, the city and New York State granted the private railroad companies the right to another valuable municipal asset: the center of the city's public streets.

Vanderbilt made a fortune during the Civil War, leasing his steamboats to the Union Army to transport troops and supplies. As the war drew to a close, he began to acquire shares of the New York & Harlem, Hudson River, and New York Central Railroads. The Central, in 1853, consolidated ten small upstate railroads built parallel to the Erie Canal, following the water-level route across the state. Vanderbilt became president of the Central and then set a plan in motion to combine it with the Hudson River Railroad, forming the New York Central & Hudson River Railroad. In a few more years, Vanderbilt had assembled a railroad

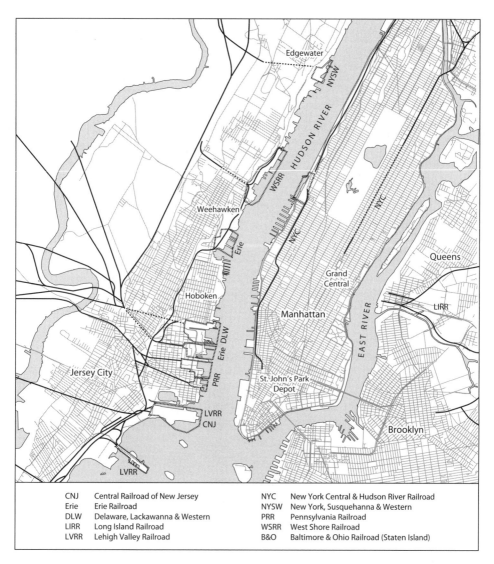

CNJ	Central Railroad of New Jersey	NYC	New York Central & Hudson River Railroad
Erie	Erie Railroad	NYSW	New York, Susquehanna & Western
DLW	Delaware, Lackawanna & Western	PRR	Pennsylvania Railroad
LIRR	Long Island Railroad	WSRR	West Shore Railroad
LVRR	Lehigh Valley Railroad	B&O	Baltimore & Ohio Railroad (Staten Island)

Map 6.1. The Port of New York railroads, showing the New York Central Railroad in Manhattan and rival railroads in New Jersey. The New York Central freight operations enjoyed an enormous advantage, with tracks running along the Hudson River to St. John's Park Depot in Lower Manhattan. All of the rival railroads' tracks ended on the New Jersey waterfront, a mile short of their final destination. *Source*: Bill Nelson map from Kurt Schlichting, *Grand Central's Engineer: William J. Wilgus and the Planning of Modern Manhattan* (Baltimore: Johns Hopkins University Press, 2013)

system from New York City to Buffalo. He then acquired other railroads, extending his system to Chicago. His railroad empire grew to be the second largest in the country, after the Pennsylvania Railroad, always a fierce rival. The Central's greatest asset remained its direct access to Manhattan Island. Vanderbilt constructed Grand Central Depot at 42nd Street, to serve as the primary passenger terminal in the city for all of his railroads until this structure was replaced in the beginning of the twentieth century by the magnificent Grand Central Terminal.[7] The Central aggressively defended its monopoly on direct access to Manhattan and the right to use the city's streets.[8]

New Jersey–Based Railroads

All of the other major railroads in the country—the Baltimore & Ohio, the Erie, and the Pennsylvania—realized that their trains must reach the Port of New York. The smaller railroads serving New Jersey and eastern Pennsylvania—the Central Railroad of New Jersey; the Lehigh Valley; and the Delaware, Lackawanna & Western—also raced to offer passenger and freight service to the port. These railroads acquired track rights and built train yards in Jersey City, Hoboken, and Weehawken. Not one of them could bring trains onto Manhattan Island. All of their rail facilities terminated on the New Jersey side of the Hudson River, a mile-wide barrier (map 6.2). Hundreds of freight cars arrived on the Jersey shore each day, along with thousands of passengers, and both needed to cross the Hudson to their final destination: Manhattan.

Thus railroads on the New Jersey side desperately required space on the Manhattan waterfront. If not, they would fall behind their rivals. Therefore, at great expense, these railroads secured permanent access to the Manhattan shorefront. With long-term leases for piers and bulkheads, the railroads built large freight pier sheds on West Street. The Delaware, Lackawanna & Western shared its freight depot at Pier 13 with the Starin Line, which operated both Hudson River and Long Island Sound steamboats.

In a brief period of time, the railroads pushed aside the maritime interests. For example, between the Battery and Cortlandt Street, the Pennsylvania Railroad leased Piers 2, 3, and 4 and the bulkhead between the piers. Up the Hudson River, the Lehigh Valley Railroad leased Pier 8, and the Central Railroad, Piers 10 and 11. Three ferries, two to New Jersey and one to Staten Island, had terminals on Piers 14 to 18. By the close of the Civil War, railroads and ferries occupied most of the waterfront along the Hudson River between the Battery and Gansevoort Street, just north of Greenwich Village.

For centuries, the Port of New York's prosperity had rested on its international and coastal shipping links with Liverpool, Europe, the cotton ports in the American South, and the city's inland waterway empire. Yet in the later fight for space, shipping companies—from the transatlantic steamships to small sloops and oyster boats—found themselves edged out, by both Vanderbilt's lines and the other railroads whose tracks terminated across the river in New Jersey. As the short-term leases held by many of the small shipping companies expired, the railroads, with far more financial resources and political muscle, negotiated long-term leases for piers on the Hudson River.

≋ The Steamship Revolution

The New York packet ships, and the clipper ships that followed during the era of the California Gold Rush, represented the culmination of the great age of sail. The Black Ball packet ship *Yorkshire* set that line's record, sailing west from Liverpool in sixteen days, an average of over 150 nautical miles a day. During the gold rush, the clipper ships set records sailing to San Francisco and on to China. The steamship revolution, however, would eclipse even the great age of sail and dramatically reduce passage times, both on the Atlantic Ocean and on the Hudson River routes to Albany and Long Island Sound.

The Industrial Revolution brought steam power to the city's waterway empire, offering faster voyages across the oceans, down the seacoast, or through the Sound to New England. Iron replaced wood as the primary construction material for ocean-going and coastal ships. With an iron skeleton and a hull built with iron plates, ships could be larger in both length and width and thus carry more freight and passengers.

The first use of steam power on ocean-going vessels involved sailing ships with an engine and paddlewheels, used as auxiliary power when the wind died. In 1817, a group of cotton merchants in Savannah, Georgia, contracted with the Flickett & Crockett boat yard on the East River to build a sailing ship with a steam engine and detachable paddlewheels.[9] On August 22, 1818, the *Savannah* slid down the ways into the East River.

In March of the following year, the steamer left New York for the run south to Savannah and reached that city in eight days. On May 24, the ship set off across the Atlantic Ocean for Liverpool, arriving twenty-seven days later. The *Savannah* steamed up the Mersey River to great fanfare, but in crossing the Atlantic, the ship mainly relied on the wind and had used its auxiliary steam engine for only 80 hours.

On both sides of the Atlantic Ocean, entrepreneurs raced to raise money to begin transatlantic steamship service. A steamship could make six or seven crossings a year, compared with three or four for the fastest packet ships. On the British side of the Atlantic, in 1839 the famous engineer and polymath Isambard Kingdom Brunel built the *Great Britain* in Bristol, England, for the Great Western Steamship Line. At 322 feet long, the *Great Britain* exceeded the length of the largest packet ship—the 216-foot *Amazon* on London's Black X Line—by over 100 feet. Built with an iron hull, not wood like the packet ships, she had a screw propeller, which was more powerful and reliable than a paddle-wheel. The true revolution came with this ship's first crossing to New York, in fourteen days! The fastest passage of any packet ship in the half century between 1818 and 1868 took sixteen days, and most crossings averaged thirty-five days, more than double the *Great Britain's* time. By crossing the Atlantic Ocean in twenty fewer days than the average packet ship, the *Great Britain* ushered in a new era that changed transportation and communications in the Atlantic World, nineteen years before the first transatlantic telegraph cable.

The success of the Black Ball and other Liverpool packet lines rested on their scheduled *departures* from New York and Liverpool, no matter what the weather, but the packets could not control the dates of their *arrivals*. While they averaged thirty-five days going across the Atlantic Ocean, passengers boarding in Liverpool had to hope the weather favored their crossing. The steamships, on the other hand, advertised *both* departure and arrival dates. New York newspapers reported when a steamship missed its scheduled arrival, something they seldom reported for packet ships. Once the steamship lines began regular service across the Atlantic, they competed to offer the fastest service, which further spurred innovations in ship design and the construction of more-powerful steam engines.

The British government offered substantial subsidies to carry the Royal Mail across the Atlantic Ocean. Samuel Cunard—who hailed from

Halifax, Nova Scotia, where he operated a successful steam-ferry business—hurried to England and, with a group of Glasgow investors, bid on the mail contract. In 1839, Cunard won the contract to carry the mail twice a month, by steamer, from Liverpool to Halifax and then on to Boston, not New York. For decades, the proud Boston merchants watched as the Port of New York grew, while the maritime commerce of their city fell further and further behind. The Cunard Line's service promised to reverse the decline of Boston's port, and the first Cunard steamships to arrive there enjoyed a warm welcome. Cunard's subsidy from the British government to carry the mail—£60,000 a year, which was the equivalent of over $300,000 in 1840—gave that line a substantial advantage over the competition. In 1848, with additional mail subsidies from the British government, Cunard added direct service to New York from Liverpool.

Once the Cunard Line began service, wealthy passengers abandoned the packet ships. Steamships soon carried the money and financial documents that underpinned the cotton web. Bills of exchange—the heart of the merchant banking system—which once were negotiated for three months' duration, could now have terms of thirty days or less, substantially reducing the risk involved in the international cotton trade.[10]

On the American side of the Atlantic, New York merchants and shippers complained loudly about Cunard's subsidy, and his steamship line would not remain without rivals. In the 1830s, the New York–based shipping firm of Israel Collins and his son Edward prospered, shipping cotton north by running a packet line between New Orleans and New York. In 1835, they loaded one of their New Orleans packets, the *Shakespeare*, with cotton and sent it to Liverpool. The ship returned to New York with textiles and manufactured goods. The profits thus earned convinced Edward Collins to start his own packet service to Liverpool, the Dramatic Line, the following year.

Collins carefully followed the success of Cunard's steamship service to Boston and New York. In response to the subsidy provided by the British government, the United States Congress awarded a mail subsidy of $200,000 to Collins, and in 1848 he began the United States Steamship Line. He immediately ordered four new wooden steamships from the East River shipyard of Brown & Bell: the *Atlantic*, the *Pacific*, the *Arctic*, and the *Baltic*. John Morrison, in his *History of American Steam Navigation*, estimates the cost of each ship at $736,000, far more expensive than a Liverpool packet ship powered by sail alone.[11]

On April 27, 1850, Collins's *Atlantic* left New York for Liverpool. Now Cunard would face stiff competition from an American line. The following year, the *Baltic* set a new record for the passage from Liverpool to New York: 9 days and 18 hours. No wonder the steamships captured the market for high-end passenger service across the Atlantic Ocean. For all of the fame of the Black Ball packet ships and their hard-driving captains, the fastest passage ever completed by a packet ship took sixteen days.

The *Baltic*'s record-setting speed came at a cost. The new steam engines, built by the Novelty Iron Works, proved to be unreliable and in need of constant repair: "The expenses at the end of every return trip to New York for repairs to the engines and boilers, after the vessels had been running a short time, was very great. Large numbers of mechanics being [sic] sent from the Novelty Iron Works, who worked day and night until the repairs to the machinery were completed, that in some cases were but a few hours before the time of sailing, that had become necessary by the heavy strain that had been put on the machinery during the voyage."[12] On one voyage, another Collins ship, the *Atlantic*, broke a drive shaft a few days after leaving and had to use its sails to return safely to New York for repairs. An army of mechanics from Novelty worked around the clock to repair the damage.[13]

Then disaster struck. On September 27, 1854, south of Cape Race at the southern tip of Newfoundland in Canada, the *Arctic*, racing through dense fog in hopes of setting another record, collided with the French ship *Vesta*. The captain of the *Arctic* turned north in a desperate attempt to reach Cape Race, but after 4 hours, the ship started to sink. Panic ensued on board. One newspaper reported that drunken passengers and crew members destroyed the bar, and liquor flowed in the scuppers along the deck. Some of the crew abandoned the passengers and seized two of the four lifeboats. In all, 35 crew members and only 14 passengers reached the coast of Newfoundland. The remaining passengers and crew—a total of 315 souls—perished, since even in late summer, the waters around Newfoundland are bitterly cold. Among the dead were Edward Collins's wife, son, and daughter.

The calamity set off a flurry of demands to improve the safety of passengers traveling on the steamships, starting with new regulations mandating a sufficient number of lifeboats aboard, with space for all passengers and crew. A *New York Times* editorial assigned the major cause of the tragedy to "an entire lack of disciplined control over the crew of the ship. No authority enforced or directed their actions. While the passengers worked the pumps, the firemen, seamen, engineer . . . seem to have been seeking their own security."[14] Captain Luce did not survive to face harsh questioning about the breakdown in discipline. Yet with only four lifeboats on the *Arctic*, even the most heroic efforts of the captain and crew could not have saved all aboard.

Edward Collins, despite his personal tragedy, continued his head-long competition with Samuel Cunard. Fate, however, dealt him another blow. Just over a year after the *Arctic* tragedy, the Collins steamship *Pacific* left Liverpool on a cold January 25, 1856. Carrying only 45 passengers and a crew of 140, the *Pacific* went "missing" and disappeared without a trace.

≈ British Steamship Companies Take Control

Two maritime disasters in the space of two years led Congress to end the mail subsidies, and the Collins line went out of business. Collins and his New York State legislative allies had faced vicious opposition to the subsidies from southern interests, who saw the grants as a blatant effort to sustain the maritime supremacy of New York at the expense of the South. From that point on, British companies completely dominated the transatlantic passenger service for the coming decades and well into the next century.[15]

In the years that followed the demise of Collins's line, transatlantic freight and passenger service continued to grow, as did the number of foreign steamship lines offering service across the Atlantic Ocean. By 1870, four British lines operated between New York and Liverpool, one provided direct service to and from France, and two German steamship companies ran between New York and the ports of Bremen and Hamburg.[16]

Not only did the European companies dominate the North Atlantic luxury passenger and high-value freight business, putting more and more ships in service each year, but the size of their ships also increased dramatically, a result of the switch from wood to iron in ship-building. This new generation of steamships, all iron-hulled, traveled at twice the speed of the fastest packet ships and, in moderate weather, sustained that speed for the entire voyage. Technological advancements revolutionized ship construction and led to an exponential increase in the power and efficiency of marine engines.

One of the first four Black Ball Line packets, the *Amity*, constructed in 1816, measured 106 feet in length, with a beam of 28 feet, and drew 14 feet of water (the depth of its hull below the waterline). The *Amity* carried 382 tons of cargo, a crucial measurement that determined the profitability of a ship, since the more tonnage that was carried, the greater the profit. By the end of the packet era, the Black Ball's *Amazon* measured 216 feet, almost exactly twice the length of the *Amity*. In comparison, at the beginning of the twentieth century, the great transatlantic passenger ships would reach over 1,000 feet in length. The steamship revolution completely changed the requirements for a modern port's maritime facilities, especially the need for much longer piers.

The Cunard Line provides a model for this revolution in ship technology. Samuel Cunard began his line running wooden-hulled, side-wheel paddle steamboats like the *Britannia*, which measured 207 feet in length, with a beam of 34 feet. The ship's steam engine produced 740 horsepower and averaged a speed of 8.5 knots.[17] The massive engine took up a great deal of space below deck, as did the coal bunkers. Cunard's first iron-hulled steamship, the *Persia*, built in 1855, measured 370 feet, 163 feet longer than the *Britannia*. Its steam engine delivered much more power—4,000 horsepower—at an average speed of 13.8 knots. In 1862, the *China*, at 335 feet long, included a key technological improvement: the elimination of the side-wheel paddles and the substitution of a screw propeller. The propeller drastically increased efficiency, thus consuming less coal, at a significant savings in operating costs.

Technological innovations accelerated after 1881, when Cunard launched its first steel-hulled ship, the *Servia*. This new ship measured 545 feet long and averaged 16.7 knots. She was capable of crossing the North Atlantic in under eight days, given the right weather conditions. In 1893, the Cunard line built twin ships, the *Campania* (fig. 6.2) and the *Lucania*, both 625 feet long. They represented the height of luxury and confirmed Britain's domination of transatlantic passenger service. The *Campania*'s twin steam engines produced 26,000 horsepower and drove the ship at an average speed of 22 knots. In May 1893, the *Campania* set a new record for crossing the North Atlantic westward: 5 days and 21 hours. Now a passenger or a shipping company could depend on scheduled service between England and New York in six days! In half a century, this revolution in ship construction changed the nature of Atlantic

Figure 6.2. The Cunard Line steamship *Campania*, 1893. This new steamship was 625 feet long, three times the length of the Black Ball Line packet *Isaac Webb*. The new generation of ships required much longer piers, which the Department of Docks had to build at great expense. *Source*: Heritage Ships

travel from crossing a hazardous, 3,000-mile ocean barrier in about forty days to a routine passage of a few days.

～～ Challenges to the Port's Infrastructure

Steamships the size of the *Campania* could not find long-enough piers on the East River, which had been the heart of the city's maritime world since the Dutch arrived. The Hudson River shoreline provided the only place where piers of sufficient length could be built to accommodate vessels as large as the new Cunard liners.[18] The increased size of the ships, however, was not the only challenge the Port of New York faced. The quasi-private, quasi-public system on the waterfront, which began in the 1600s with the first water-lots, had created a hodgepodge infrastructure. Piers differed in length and width, and most did not include sheds or other structures to provide shelter for passengers and cargo. Few owners kept the river bottom dredged to the depth needed for heavier ships, or their wharves and piers in good repair.

The city and state of New York exercised little regulatory power. The city's water-lot grants had transferred the responsibility for the port's infrastructure to private owners. No single entity exercised control over the Manhattan waterfront, the surrounding waters, or the port as a whole. A *New York Times* editorial in 1857 emphasized the city's need

Figure 6.3. South Street, on the East River, looking south from the Brooklyn Bridge toward the Battery. The bowsprits of the packet ships reached across the congested street. Counting-houses and warehouses crowded the adjacent streets in 1825. *Source*: Edward W. C. Arnold Collection of New York Prints, Maps and Pictures, Metropolitan Museum of Art

for an efficient, functioning harbor, since "upon this commerce the prosperity of our metropolis and all of the adjacent country depends."[19] As the foreign-based steamship lines arrived, they competed for space on the waterfront with coastal packets, Albany and Long Island Sound steamers, Erie Canal boats, ferries, and the powerful railroads.

Along the East River, the piers had been of adequate length to serve the first generation of packet ships. Vessels like the *Amity* would tie up to a pier with their bowsprits extending over South Street, pointing to the counting-houses across the street (fig. 6.3). While South Street provided illustrators and marine artists with romantic images of sailing ships and a waterfront atmosphere, the crowded piers and streets, with cargo strewn all over, symbolized the inefficiency and obsolescence of the East River waterfront.

The piers along South Street, constructed by the water-lot grantees, varied in length and width. Piers 27, 28, and 29 all extended over 400 feet out into the East River and measured between 33 and 40 feet in width (map 6.2). By comparison, Pier 36, at a length of 305 feet, could accommodate ships measuring only less than 300 feet, not the Cunard's longer liners, such as the *Persia* (370 feet) or the *China* (335 feet). The

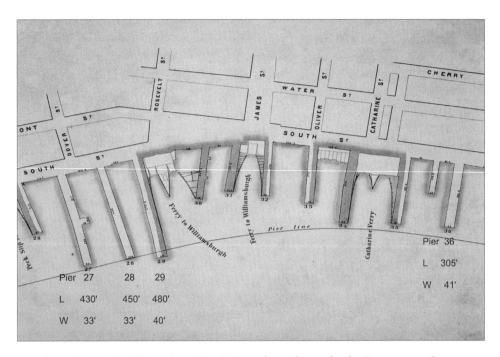

Map 6.2. The East River piers, 1860. Prior to their takeover by the Department of Docks, the piers along the East River were a hodgepodge of different lengths and widths. Few had sheds to store cargo, and their private owners did not properly maintain the piers or keep the waterways dredged. On the map, the length (L) and width (W) of the piers is given in feet. *Source*: Created by Kurt Schlichting. Source GIS layer: historic pier map, New York Public Library, Map Warper.

new generation of passenger and freight steamships had no choice but to move to facilities on the Hudson River.

The question of extending piers of the necessary length out into the Hudson, in order to accommodate the new ships, created controversy. In 1837, the State of New York established virtual pier lines around Manhattan, beyond which piers could not be built (see the line on map 6.2). A number of the East River piers, already extending to the pier line, could not be lengthened farther. With no room to build longer piers along the East River, the shipping industry turned to the Hudson River, and the center of the maritime industry shifted to the West Side of Manhattan Island. The state's pier lines on the Hudson, however, still did not allow for piers long enough to secure the transatlantic steamships, often 800 or 900 feet long.

Development along the Hudson shorefront followed the same pattern as on the East River: a haphazard, unplanned combination of private water-lots and pier construction, which created a jumble of private and city-owned wharves and piers. The *Times* editorial did not exaggerate, as the chaos along the shore *did* threaten the maritime commerce of the city and the entire port, since shipping companies and the European steamship lines had alternatives. They could move their facilities across the Hudson River to New Jersey, where Jersey City and Hoboken waited with open arms. To add to the threat, the cities of Baltimore, Boston, and Philadelphia never ceased their efforts to persuade, cajole, or bribe, if necessary, the maritime companies to relocate from the Port of New York.

Maintenance was neglected along both of Manhattan's shorelines. Piers literally fell apart, and the slip space between piers, silted up with debris and sewage, was not dredged. Venal harbor- and dockmasters, appointed by the state to manage the waterfront, demanded bribes to obtain dock space, and criminal gangs stole freight at all hours of the day and night. Chaos reigned. A *New York Times* editorial proclaimed: "Abuses of every kind exist in this harbor, whose aggregate is imposing a heavy tax upon all ships and merchandise seeking this port. Prominent among them are the abuses connected with the piers and docks of the City."[20] The *Times* warned that without sheds and warehouses, cargo left on the open piers is "exposed to the depredation of thieves and pirates who swarm like locusts," stealing over $1 million of merchandise a year.

An article in *Harper's New Monthly Magazine* described the heroic efforts of the New York City Harbor Police to battle the "dock rats": gangs of young thieves who lived under the piers and preyed on the maritime commerce.[21] Junk shops on South and West Streets fenced these stolen goods. Organized gangs smuggled goods from ships at night, to evade import taxes. With few officers and only one steam-powered patrol boat, the police were waging a losing battle to control crime on the waterfront, a struggle that continues to the present day.[22]

City merchants, foreign shipping interests, owners of coastal packet lines, captains of small schooners, and ferry companies complained and demanded change. To add insult to injury, the railroads now controlled a huge part of the waterfront. The 1857 editorial in the *Times* captured this anger and frustration: "Lawlessness and Filth reign supreme on our wharves and every cargo landed on the foul and dilapidated piers which fringe the East and North [Hudson] Rivers is more or less injured by these causes. The slime of the City sewers and the wash of ever-filthy streets of New York and Brooklyn is deposited in every dock shoaling the water."[23] By the mid-1850s, New York City's merchants, shipping companies, mariners, politicians, civic leaders, and press all realized that the increased crowding, chaos, and deterioration along the Manhattan waterfront posed a dire threat to the very source of the city's wealth.

≈ Reform on the Manhattan Waterfront

Eventually the politicians in New York City and the New York State Legislature could no longer ignore these myriad problems. In 1855, the legislature passed an act "for the appointment of a commission for the preservation of the harbor of New York from encroachment and to prevent obstructions to the necessary navigation thereof."[24] The powerful New York Chamber of Commerce, whose members included the prominent merchants and shipping companies in the city, supported the commission. The ever-corrupt Democratic Tammany machine, which benefited from the chaos, corruption, and status quo on the waterfront, resisted.

The act directed the commissioners to recommend new regulations limiting the construction of piers farther out into the rivers, and the commissioners also devoted a great deal of energy to the problems on the shoreline: the piers, wharves, and bulkheads. Over the next two years, the commissioners took testimony from forty-seven witnesses, including private pier owners, wharfingers (individuals who managed piers either for private owners or the city of New York), harbormasters, merchants, shipowners, and longshoremen.

The commissioners pinpointed the central problem: the incredibly complicated division of ownership and responsibility along the Manhattan shoreline. They traced the problem back to the system of water-lot grants that transferred ownership of this real estate from the city of New York to private individuals. Although the water-lot grants required private owners to build and maintain their bulkheads and piers, the city had little real power to ensure that the piers were properly constructed and maintained. No single city department had direct oversight of the waterfront, and the absence of governmental responsibility and enforcement led to the crisis: "[Regarding] the enforcement of the laws for the government of the harbor . . . under the present system, this duty, in the City of New York, is committed to the superintendent of wharves and slips, to the harbor and dock masters, and to the police and is performed by none. The wharves remain uncleaned. . . . They are in bad repair. . . . In short every regulation which has been established, whether for the preservation of the harbor, or for the protection of the merchant or the wharf owner, is disregarded, and remains a dead letter, from the absence of a supervisory and controlling power."[25]

S. A. Frost, a wharfinger for twenty-four years, managed a number of piers for private owners: Piers 16, 17, and one-half of 18 on the East River, plus 447 feet of bulkhead. Frost testified that the piers cost $93,606 to build and that wharf fees, all from transient ships, totaled about $11,000 a year. After costs of $2,794, this generated a net return of 8.8 percent.[26] The private owners of the piers persistently complained that the then current fees did not generate an adequate return and continuously lobbied the state to raise them significantly, but the commissioners' final recommendations did not include an increase. On the contrary, they compared the port fees in New York with those in Baltimore, Boston, and Philadelphia and found that these were higher in the latter three cities. Raising its fees would undercut the Port of New York's cost advantage.

Other testimony documented the deplorable condition of the piers and bulkheads, including some that had simply collapsed into the water. Few piers had covered pier sheds. Cargo, once unloaded, if not carted away immediately, simply sat on the docks in all weather, causing significant damage to it. Filth of all sorts accumulated between the piers, requiring repeated dredging that the private owners delayed as long as possible. Isaac Orr, who was engaged in dredging operations in New York Harbor, reported that the average accumulation of deposits of filth and garbage in the slips on the Hudson River to be 5 feet every three years, and along the East River, "about a foot per annum."[27] Wharfinger S. A. Frost added that the filth required dredging every two years, because of deposits, "principally from the sewers and the wash of the streets. Another cause is from dumping rubbish from houses and factories into the slips, at night. Rubbish is also thrown in from vessels."[28] Frost also testified that the harbormasters, who assigned arriving ships to an open dock, demanded bribes to let them dock quickly:

Q. Do you know if unauthorized charges are made by public officials for obtaining berths?

A. I have frequently had complaints from captains of vessels of such charges being made. . . .

Q. Is the practice you refer to well known and notorious?

A. It is.[29]

The commissioners' report documented the corruption surrounding the harbormasters and led to further investigation. In 1862, the New York State Senate established a committee to "examine into the affairs, and investigate the charges of malfeasance in office of the harbor masters of New York."[30] Witnesses called to testify suffered from what the commissioners termed "know-nothing-ism" and refused to provide details about the bribery on the waterfront. Other testimony focused on theft and crime. All of the piers hired night watchmen, but the thieves often outsmarted them, threatened violence, or offered sufficient bribes for the watchmen to look the other way. Shipping interests in the port faced not only deteriorating piers, overcrowding, and delays, but they also had to deal with serious corruption and crime.

The 1857 commissioners' report pointed to the dramatic differences between the ports of New York, Liverpool, and London regarding the costs associated with creating the facilities necessary to accommodate the world's shipping. In both London and Liverpool, the local governments invested a fortune to build wet basins along the shore. In London, the Thames River waterfront could not accommodate large docks, so, at great expense, excavations were carried out to create wet basins and then build docks and warehouses to enclose each basin with "massive walls of hewn stone, and fitted with every appliance for the protection and speedy lading [loading] and discharge of cargo . . . [that] comprise one thousand seventy-one acres varying for sixteen to twenty-five feet in depth, and costing sixty-five millions of dollars."[31] In Liverpool, the government constructed 174 wet basins, creating 14 miles of enclosed wharves at an estimated cost of more than $200 million, the equivalent of $6 billion today.

By comparison, in 1857 the city of New York owned eighteen piers on the East River and twenty-six along the Hudson, whose construction costs amounted to just over $1.8 million, a small fraction of the amount incurred by Liverpool and London.[32] Along the Manhattan shoreline, a bulkhead at the high-water line and relatively simple wooden piers built out into the rivers sufficed. New York Harbor's unparalleled topography allowed the construction of a maritime infrastructure at minimum expense. The private water-lot grantees had invested as little as possible and earned handsome returns. Collectively, they saw little reason to increase their investment.

The commissioners considered a number of reform proposals, including raising wharf fees and adding a duty on the landed cargo, to provide additional income to be used to repair and, in some cases, rebuild the piers entirely. After two years of work, the commission decided to not recommend raising fees for the use of the piers, since the threat of shipping companies moving to Boston, Philadelphia, or across the Hudson River to New Jersey loomed. Adding a duty on incoming cargo, to be paid by the merchants, also generated strong opposition.

Instead, the commission focused on the lack of clear responsibility for overall management of the port and the waterfront. To solve this problem, the report recommended that New York State establish a "board of commissioners with exclusive powers over the harbor within the jurisdiction of this State."[33] Three commissioners would be appointed by the governor and the New York State Senate, and the mayors of New York City and Brooklyn would serve as two additional commissioners. The clear intent of this recommendation was to replace the private/public sharing of power over Manhattan's waterfront, going back more than 200 years, with an authority responsible to the state government, not to the city of New York alone.

To no one's surprise, the mayor, the Common Council, and Tammany Hall opposed any plan to shift local control to the state, despite the seriousness of the existing problems and the threat that maritime commerce would leave the port. An overall authority for the port, independent of the seemingly intractable political corruption in the city, would have to wait another sixty-four years, until the formation of the Port Authority of New York and New Jersey in 1921. In the short run, no serious efforts to improve the port facilities followed after the commissioners released their report. If anything, the waterfront became more crowded and, at the same time, more dilapidated in the decade before the Civil War.

Reform in the Port of New York was further delayed by the Civil War. In 1867, the commissioners of the Sinking Fund undertook a major study of the city-owned piers. This fund was established in 1817 by New York City's new charter, in order to manage the city's indebtedness. To avoid political corruption, the fund used dedicated revenues to pay the city's outstanding bonds and ensure its creditworthiness. Fees from water-lot leases went directly to the Sinking Fund, and the commissioners supervised the city-owned piers and bulkheads.

The Sinking Fund's commissioners completed a detailed study of "the conditions of the wharves, piers, and slips belonging to the city," including estimates of their present value, the amount required to rebuild them, and their value when repaired.[34] City engineers surveyed every city pier and adjoining bulkhead. For each, they included a detailed drawing of both the pier and the bulkhead, with their dimensions and the depth at low water. This exhaustive two-volume study represents the first time any civic body had compiled a detailed mapping of sections of the Manhattan waterfront's infrastructure.

For example, on the Hudson waterfront, Pier 33, at Jay Street, was 523 feet in length and stretched beyond the pier line established by the state as the outer boundary for any pier along the Hudson River. The report described this pier as being in "poor condition." An extension on the right side needed to be removed. At the outer end of the pier, the 21-foot depth at low water could accommodate large ships, but along the pier the depths reached only 9 feet, which required dredging. West Street had no water at low tide, so even the smallest ships or barges could not tie up to the bulkhead. The report estimated the then current value of the Jay Street pier at just $50,000, given its dilapidated condition, and it required $25,000 in repairs, including dredging. Once repaired, the pier's value would increase to $125,000, and the city would have a functioning 500-foot pier to accommodate larger ships using the port. On the East River, Pier 29, at the foot of Roosevelt Street, adjacent to the Williamsburg Ferry terminal (see map 6.2), needed to be completely rebuilt. The outer section of the pier had collapsed into the river and could not be used. The deplorable condition of this city-owned pier mirrored the condition of most of the piers and bulkheads along the Manhattan shorefront.

The estimated cost to repair and rebuild just the city-owned piers totaled $1,119,685. To extend all the piers out to the pier line added another $791,550, for an overall total of $1,911,235. When repaired and extended, the value of the piers and bulkheads would increase to over $20 million from their then current $15 million.[35] The city, however, faced a daunting problem: how to finance the needed work. In 1867, the entire budget for the city of New York totaled $5 million.[36] The estimated amount required for rebuilding the waterfront would be the equivalent of 38 percent of all spending in 1867. In the context of New York City's 2016 budget of $75 billion, 38 percent amounts to $28 billion. Moreover, the cost estimates did not include upgrading the hundreds of private piers and bulkheads. Where would the city and the private owners find such staggering sums?

If the thousands of Erie Canal boats could not find space in New York's harbor, the Boston & Albany Railroad stood ready to load the bounty of the Midwest onto freight cars in Albany and deliver it to the port of Boston. Across the Hudson River, only a short ferry ride away, Jersey City and Hoboken eagerly waited for the transatlantic steamship companies to relocate to their piers. (In the 1860s, the Cunard Line did move to Jersey City for a brief period of time.) The Pennsylvania Railroad threatened to divert its huge freight traffic from New York to Philadelphia if the transatlantic and coastal shipping companies moved their operations to the latter city's piers on the Delaware River.

The 250-year evolution of the hybrid system of private/public control of the waterfront on Manhattan Island had failed to keep pace with the inconceivable growth of the Port of New York and its maritime industry. A solution had to be found, or the dominance of the port and the economic vitality of New York City and the entire port region would be at risk.

[7]

The Public and Control of the Waterfront

I n the aftermath of the state investigations, political pressure for radical change in the Port of New York could not be ignored. The mayor, Common Council, and the Democratic political machine, however, opposed all efforts to move control of the port from the city to the state. Finally, in 1870, the city and the state agreed to modify the city charter and created a new municipal agency, the Department of Docks.[1] The charter gave the new department sweeping power over the Manhattan waterfront. Where private development formerly had added new land to the shoreline, built piers and bulkheads, and managed the city's piers, a public agency would now take over and eliminate the chaos on the waterfront. No longer would there be a lack of responsibility and accountability, since a public agency—not private interests—would control the waterfront.

Initially, five commissioners, appointed by the mayor, would oversee the new department, but in 1873, the state reduced that number to three. The legislation granted far-reaching powers to the Department of Docks. Sidney Hoag, assistant chief engineer of the department in 1905, details these extensive changes, which gave local government broad control over the city's premier asset: the water surrounding Manhattan Island. The Department of Docks had jurisdiction over the entire Manhattan waterfront, including Governors Island, Ward Island, Randall's Island and Blackwell's Island (now Roosevelt Island)—a total of 38 miles of waterfront. The department was able to

- take possession of private piers and bulkheads, with owner agreement as to the purchase price;

- request the City Council to take legal proceedings to acquire piers and bulkheads by condemnation, if a price could not be agreed on;
- make contracts for constructing and repairing piers and for dredging slips;
- build pier sheds, for the protection of cargo;
- lease, at public auction, piers and wharf property belonging to the city for a period not to exceed ten years; and
- make rules and regulations for the use of the piers and bulkheads.[2]

At first, the new department would have a limit of $350,000 a year to spend on repairs to and reconstruction of the piers, bulkheads, and slips. In addition, the city comptroller would issue bonds, with the approval of the commissioners of the Sinking Fund, to repurchase the waterfront. The amount of bonds to be issued could not exceed $3,000,000 each year.

The new charter explicitly separated the finances of the Department of Docks from the general budget of the city. Leases for piers and ferries, as well as wharfage fees for the short-term use of piers and bulkheads, would be collected by the Department of Docks and kept in a separate account. The monies were to be used to cover expenses and pay the interest and principal on the bonds. Any remaining funds would be transferred to New York City each year, to reduce city debt. The charter prohibited the use of any city tax revenues for the Department of Docks, so it would be self-financing. To the modern ear, an expectation that a governmental agency would be self-financing from user fees and not rely on taxpayers may seem discordant. Despite the dilapidated condition of the city-owned piers, documented by the Sinking Fund study, in 1856 the East River piers still generated revenues of $61,005, and those on the Hudson, $65,500. Once the department repurchased, rebuilt, and modernized the waterfront, dramatically increased revenue seemed assured.

The Department of Docks, as a public agency, exercised power independent of direct oversight by the mayor or the Common Council. The new agency's expenditures remained outside of the city's normal review and approval process. The independent structure of the Department of Docks foreshadowed the creation of public authorities fifty years later, such as the Port Authority of New York and New Jersey.

Numerous other authorities followed the establishment of the Port Authority, and, over time, they have exercised enormous power over transportation in the entire New York metropolitan region. Today a new generation of public and semipublic authorities have led the extraordinary rebirth of the city's waterfront that dazzles New Yorkers and visitors alike: the Battery Park Housing Authority, the Hudson River Park Trust, and Friends of the High Line.

〜〜 Master Plan

As soon as the new commissioners of the Department of Docks convened, they took up the task of selecting the engineer-in-chief. After some deliberation, they chose George B. McClellan, a West Point graduate, Civil War general, and the youngest candidate ever to run for president of the United States. The arrogant McClellan demanded and received a salary of $20,000 a year, twice that of the governor of New York. McClellan hired a professional staff and work began. Before any rebuilding could commence, however, the new department needed to develop a master plan for the revitalization of the Manhattan waterfront. For the first time in the city's history, a rational planning process involving the shoreline of the island commenced.

One problem arose immediately: no systematic records of the waterfront's development could be found. The department's staff members conducted a title search of all water-lot grants, going back to 1684, to try to determine who owned the waterfront. They could not find any details of when the city-owned piers had been constructed, who constructed them, and at what cost. Even more-complicated problems involved the present ownership of the private bulkheads and piers, which had been sold and then resold over time. The Department of Docks planned to repurchase the privately owned piers and bulkheads and thus needed to identify the legal owners, in order to negotiate with them.

The department's engineers concentrated their initial efforts on the bulkhead, the boundary between the city's outermost streets and the water. The master plan included building a solid bulkhead wall of uniform design and height around the perimeter of the entire southern half of Manhattan Island.[3] With a solid bulkhead, and the water dredged to a proper depth, South and West Streets would serve as wharves for smaller boats and barges, adding significant space along the waterfront.

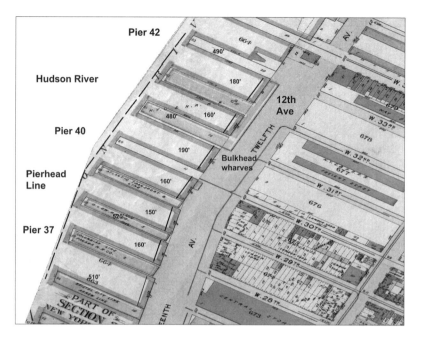

Map 7.1. The Department of Docks' master plan for a new bulkhead and piers on the Hudson River, 1891. When completed, the master plan transformed the waterfront. The new piers extended out into the Hudson River to the pierhead line and were close to or over 500 feet in length. The relatively uniform width between the piers opened up space along the new bulkheads for small ships, barges, and car floats. *Source*: Created by Kurt Schlichting. Source GIS layer: historic street and pier maps, New York Public Library, Map Warper.

The pier line, established by the State of New York in 1856, defined the maximum length for all the new piers the Department of Docks planned to build. The master plan specified uniform distances between the new piers, to ensure slip space for two large ships at adjoining piers. An 1891 map of the rebuilt Chelsea waterfront between West 28th and West 33rd Streets illustrates the successful implementation of the master plan (map 7.1).

≈≈ Reclaiming the Waterfront

Miles of piers and their adjacent bulkhead remained in private hands, a legacy of the water-lot grants, which went back two centuries. As Hoag notes, "It appears safe to estimate that the city relinquished 95 percent of the waterfront on both rivers below Forty-second Street by this practice [water-lot grants]."[4] The powerful railroads controlled much of the waterfront at that point, and they willingly paid very high lease rates for access to Manhattan's shore. The Department of Docks raised capital to buy back the waterfront by issuing bonds over the next three decades, from 1870 to 1900. It would pay $19,319,981 to do so, an enormous expenditure.[5]

The Hudson River shoreline in Greenwich Village provides an example of the complicated process to repurchase the waterfront (map 7.2). In the first two decades of the nineteenth century, on the three blocks between Houston and Morton Streets, the city granted ten water-lots to the west of Greenwich Street, then the water's edge. In turn, the grantees built West Street and then filled in the land from Greenwich Street out to the new shoreline. Among the grantees, John Jacob Astor stands out. Astor achieved fame as the founder of the American Fur Company, which monopolized the fur trade. A greater share by far of Astor's vast fortune, however, came from real estate speculation in New York, including his water-lot grants on the Hudson River.

In the case of these ten grants, over time the original grantees sold their water-lots. By the 1870s, nineteen private owners controlled this section of the waterfront. The Department of Docks could not reach an agreement with the owners, so it went to court to condemn these properties and have the court determine the final cost. The court typically awarded $500 to $600 a foot for a bulkhead, but for the 200 feet of bulkhead between Clarkson and Leroy Streets, the court awarded $203,563—significantly more than $500 a foot.[6] The department, in its haste to begin construction, seized the property and started work while the court deliberated. This action proved to be a costly move, as the court awarded the owners interest for the time between when the department started work and the final court settlements.

On the East River, the Department of Docks purchased the bulkhead between Beekman Street, Pecks Slip, and the Williamsburg Ferry. As it did on the Hudson River waterfront, the master plan included building a new bulkhead along South Street, on the river, and widening the street. From the new bulkhead line, the department planned to remove all the old piers and replace them with new piers of uniform design. The old bulkhead and piers belonged to the descendants of some of the original Dutch settlers, the Schermerhorns and Ver Plancks, who

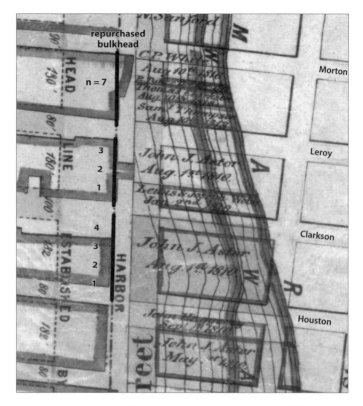

Map 7.2. Hudson River water-lots and Department of Docks bulkhead repurchase, extending from Houston Street to Morton Street. To rebuild the bulkhead, the Department of Docks had to repurchase the water-lot land on the shoreline. The made-land across West Street remained as private property, creating enormous wealth for the original grantees and their descendants. From Houston Street to Clarkson Street (four separate parcels of land), the grantee was John Jacob Astor (August 1, 1810). From Clarkson Street to Leroy Street (3 separate parcels of land), the grantees were Lewis and Joseph Webb (January 2, 1810) and John Jacob Astor (August 1, 1810). From Leroy Street to Morton Street, the grantees were C. P. White et al. (1812–1816). *Source*: Created by Kurt Schlichting. Source GIS layer: historic water-lots map, New York Public Library, Map Warper.

were grantees of the earliest water-lots. In 1905, the Schermerhorns received payments totaling $106,585, and the Ver Plancks, $323,764, close to a third of a million dollars.[7] The families also retained title to the made-land created in earlier centuries on their water-lots between Pearl and South Streets, further enriching both dynasties.

When a government condemns private property for a public purpose, the owners often argue in court that the process represents an unconstitutional "taking" of property, which is prohibited by the Fifth Amendment. The Department of Docks, however, did not buy back the water-lots, that is, the made-land out from the original shore. They typically bought *only the bulkhead* along West and South Streets. New land created by filling in the original shoreline remained the private property of the water-lot grantees and their descendants, and many great real estate fortunes in New York history can be directly traced back to the creation of new waterfront land through water-lot grants.

Today, on the block between Clarkson and Leroy Streets at West and Washington Streets, the seven properties there are assessed at a total of $10,211,100. On West Street between Leroy and Morton Streets, two sleek new apartment buildings, with stunning water views, together are assessed at $55,396,830.[8] The extraordinary value of these buildings rests on their expansive view of the Hudson, where once-decrepit piers and the crumbling West Side Highway had blocked even a glimpse of the river.

～～ Rebuilding Manhattan's Waterfront

The Department of Docks eventually bought back 17.8 miles of Manhattan's waterfront and created a new shoreline. The department first undertook a detailed study of the existing bulkheads, using underwater surveys and borings to determine how far below solid rock lay. Conditions varied, and the engineers utilized a number of designs for the new bulkhead. To lower the cost, the engineers used pre-cast concrete blocks below the water line and capped the last 10 feet with granite blocks. With the bulkhead in place, new piers replaced the rotten, dilapidated piers that threated to destroy the city's maritime commerce. Today, thousands of people enjoy the new Hudson River Park's pedestrian walkway, on the river's edge, that sits atop the granite-capped bulkhead built over 100 years ago.

A typical section of the bulkhead measured 36 feet in depth at low water. The granite and pre-cast concrete blocks rested on bedrock. Behind the new bulkhead, stone riprap fill created a solid foundation for the newly widened West and South Streets (fig. 7.1). The department set up construction yards along the Hudson River at Gansevoort Street,

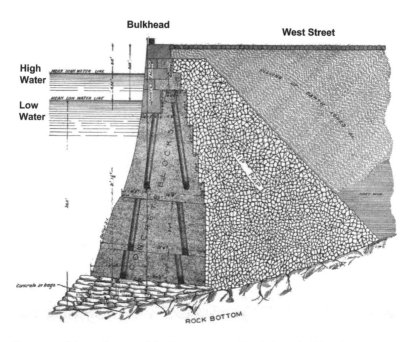

Bulkhead

West Street

High Water

Low Water

Figure 7.1. Schema for one of the Department of Docks' new bulkheads. These uniform bulkheads stabilized the Manhattan shoreline and provided a foundation for the widened South and West Streets. The bulkheads were used by the railroad's car floats and smaller ships. The newly created parks, walkways, and bike paths on the Manhattan shoreline today rest on these bulkheads, constructed at great cost by the Department of Docks more than 100 years ago. *Source*: Sidney Hoag Jr., "The Dock Department and the New York Docks," paper no. 16, presented March 22, 1905, *Municipal Engineers of the City of New York, Proceedings for 1905*, plate 36

and on the East River at East 17th Street, to build the cast concrete blocks, using molds to speed production. The concrete blocks were then loaded onto barges and towed to the section under construction. Dredges removed the debris and muck down to the bedrock, and divers, using primitive equipment, arranged concrete bags to provide a solid foundation for the first layer of cast blocks.

The heavy cast concrete blocks and granite caps needed to be moved and carefully fitted. This was done with the assistance of a giant floating derrick to lift the blocks, which weighed between 25 and 45 tons each.[9] The derrick, named the "City of New York," could hoist 100 tons. It became a familiar sight, symbolizing the rebuilding of the waterfront (fig. 7.2).

Figure 7.2. The "City of New York," a 100-ton floating derrick, lifting a pre-cast concrete bulkhead block. This innovation, the largest floating derrick in the world, allowed the Department of Docks to use such blocks to build the new bulkheads. Multiple concrete blocks could be quickly poured in construction yards on the waterfront. The blocks were loaded onto barges and towed to the section of the shoreline under construction, where they were lowered into place by the derrick. *Source*: Sidney Hoag Jr., "The Dock Department and the New York Docks," paper no. 16, presented March 22, 1905, *Municipal Engineers of the City of New York, Proceedings for 1905*, plate 21

When the Department of Docks was first organized in 1870, it contracted out the construction of the new bulkheads and piers. Difficulties with the contractors arose, however, and by 1880, the department hired its own workforce. Tammany Hall supported the change and used its political muscle to ensure that the new labor force included many of its loyalists. Scandal followed, as critics charged that the new workers had to pay kickbacks to Tammany to secure a job. The whiff of corruption lingered, but after initial difficulties, the department pushed ahead. For the next twenty plus years, work continued. Each year the engineer-in-chief submitted details of all construction activity. The 1903 report listed the completion of 7.8 miles (30,575 feet) of new bulk-head; 214 piers, with a total of 7.6 million square feet of space; and the grading and paving of 265,700 square feet of the widened West and South Streets.[10] The department had completely rebuilt the waterfront around the southern portion of Manhattan Island, creating modern facilities to sustain the city's maritime commerce.

∼∼ Increased Revenue

Revenue from both the new dock and ferry leases and from wharfage increased dramatically. This totaled $460,164 in 1871 and, with the leasing of the new bulkhead and piers, reached $2,775,172 in 1900, an increase of slightly over 500 percent. In most years, revenue exceeded expenses, generating a surplus. The Department of Docks not only met the legal requirement to be self-financing, but it also sent a significant amount of surplus funds to the city of New York to reduce the general debt, as the 1871 charter required.[11] The increased revenue covered the cost of building longer and more-modern piers along the Hudson River. By the turn of the century, the Department of Docks had spent over $11 million to repurchase the Manhattan waterfront, enabling the construction of a new, modern, maritime infrastructure.[12]

The success of the Department of Docks served the interests of Tammany Hall. As the rebuilding of the port's infrastructure expanded, the department's payroll became a source of patronage for the Democratic machine, dominated by Irish politicians. James Fisher, in his masterful work, *On the Irish Waterfront*, characterizes the department's payroll as a "patronage windfall."[13]

The department's 1887 annual report included five pages of small print listing all of the employees and either their salaries or their hourly wages. Engineer-in-Chief George S. Greene earned an annual salary of $6,000, while ninety-one laborers earned 23 cents per hour: from approximately $9 to up to $12 a week for 40–50 hours on the docks. These workers included Thomas Ahearn, Peter Burke, James Kennedy, John McSorley, John Murphy, and Thomas Sullivan.[14] Securing a job at the Department of Docks required the support of Tammany Hall. On the Irish waterfront, employment required a note or a word from the local ward boss. Tammany expected the higher-paid salaried workers to hand over 5 or 10 percent of their income as a "donation" to the machine.

Tammany corruption was pervasive in the Department of Docks. In addition to jobs, the Democratic machine also expected that the major contracts awarded by the department for building piers, dredging, and supplying construction materials would go to the politically connected. In 1884, the Union Dredging Company, which had strong ties to Tammany Hall, was paid $111,509, over 25 percent of the entire construction account for that year. Charges of corruption in the Department of Docks led to a series of investigations, which charged that Union Dredging submitted bids with no price filled in. The department's treasurer then wrote in a price, at slightly less than the lowest bidder. J. Sergeant Cram, an ex-commissioner of the Department of Docks who served in the mid-1880s, admitted that "Tammany men always had the preference, other things being equal."[15]

A decade later, with the corruption and influence of Tammany on the city's government continuing, the New York State Assembly appointed a special committee in 1894 to investigate the "offices and departments of the city of New York." Chaired by Robert Mazet, a Republican assemblyman, the committee began conducting hearings.[16] In August 1899, the committee called Charles F. Murphy, treasurer of the Department of Docks, to testify. The investigator asked Murphy to explain why seventeen dredging jobs were given to one company without competitive bidding, all on "treasurer's orders."[17] These contracts went to the Morris & Cummings Company, whose president, George Leary, served as a high-ranking officer in Tammany Hall. Morris & Cummings bought all the piles for the new piers from Naughton & Company, owned by Daniel Naughton, a "Tammany leader," according to the *New York Times*. Political connections obviously ruled the department. As Murphy's testimony before the committee admitted, "he paid a good deal of attention to the patronage of the department. The men were mostly sent to him by [Tammany] district leaders he said. . . . He said he looked on that as a proper way of selecting appointees and a legitimate method of rewarding proper party service."[18]

The influential New York Chamber of Commerce also publically criticized the department's overall management of the waterfront in its annual report for 1896, listing "certain ills":

1. An absolute lack of proper wharves and docks.
2. A most exorbitant charge for the use of the piers that do exist.
3. Steamship lines are also compelled to pay for the dredging of the docks.[19]

A. Foster Higgins, chair of the chamber's Committee on the Harbor and Shipping of New York, demanded that the city of New York close the Department of Docks and turn the piers back to private companies.[20] Higgins's argument, however, ignored the waterfront's history. Private interests, which had dominated the Manhattan waterfront from the colonial era to the middle of the nineteenth century, had completely failed to keep pace with the ever-increasing maritime commerce of the port. Despite the corruption, the comparative success of the Department of Docks proved that only a public enterprise, responsible for the entire waterfront, could marshal the financial resources and political will to improve the port's infrastructure.

Nonetheless, the Chamber of Commerce did correctly identify one of the most serious problems on the rebuilt waterfront: the intense competition for space among the major users of the port. Of the 13,439 feet of space on the Hudson waterfront from the Battery to Gansevoort Street, foreign steamship lines occupied only 1,779 feet (13.2%) and coastal shipping, 1,861 feet (13.9%). The railroads took up more space than either of these two major shipping businesses combined: 3,883 feet (28.9%).[21] Without the ability to bring freight back and forth to Manhattan, they could not compete with the only railroad with direct access to the island: their bitter rival, the New York Central.

Despite the legislative investigations into political corruption and criticism from the city's chamber of commerce, at the beginning of the new century, the future looked bright for the port, the Department of Docks, and the port's maritime commerce. New York reigned as the busiest port in the world, with—in terms of both imports and exports—no serious American rival. The city also remained the leading manufacturing center in the country, and its population continued to grow. In 1898, Manhattan and the four outer boroughs—Brooklyn, the Bronx, Queens, and Staten Island—merged to form greater New York City, the most populous place in the United States. Across the river in New Jersey, the cities along the Hudson River also prospered as their maritime and railroad-based commercial activity increased. The Port of New York now encompassed not just the Manhattan waterfront, but over 650 miles of shorefront in the bi-state metropolitan region, seemingly enough shoreline to meet the demands of the twentieth century.[22]

≈ The Need for Longer Piers

During its first two decades, the Department of Docks devoted a great deal of energy to building the bulkhead around the lower part of Manhattan Island, acquiring the decrepit docks owned by the water-lot grantees, and constructing modern piers. All were major accomplishments, but a serious problem remained: the ever-increasing size of ships, especially the British transatlantic steamships that necessitated longer piers.

The Montgomerie Charter granted control of the waters around Manhattan Island out to 400 feet beyond the low-water line to the city of New York. Well into the nineteenth century, piers built out into the East and Hudson Rivers did not extend that far. As the dimensions of ships increased, however, piers on the East River could not be lengthened, since the construction of new, longer piers could be accommodated only along the Hudson River, where a mile of water separated the Manhattan shoreline from New Jersey. In 1856, New York State had asserted its legal control of all waters beyond the 400-foot mark. At that point, the state's New York Harbor Commission established permanent pier lines around the entire island, beyond which no piers could be built. To further complicate the issue, in 1890 the US Congress passed legislation preempting the State of New York's control of the waterways in the port and delegated responsibility to the US Army Corps of Engineers. Now, all plans to build beyond the newly established pier line required permission from the Corps of Engineers, representing the federal government. The government wanted no interference in the waterways around the country, in order to allow US Navy ships to maneuver.

With increasing numbers of transatlantic steamships coming to the port, the Department of Docks desperately needed to build longer piers that extended beyond the established pier line. The department petitioned for exemptions, and the Army's Corps of Engineers refused the request. Only one solution proved to be feasible. Since piers could not extend farther out into the Hudson River, the Department would have to repurchase the made-land on the waterfront in Greenwich Village and, later, in Chelsea; excavate; and move the bulkhead 200 feet back toward Manhattan's original shoreline! Here the repurchase included not just the bulkhead, but the actual made-land created by filling in the water-lots a century earlier. Without the longer piers, the transatlantic steamship companies would have to abandon the Port of New York.

The construction of the longer piers required an enormous expenditure, one beyond the capability of the private pier owners. Only a public agency, with the ability to issue bonds, could undertake a project on this scale. The Department of Docks, as part of the New York City government, could use eminent domain to acquire the made-land along the Hudson River from its private owners. As in Liverpool, the department would undertake a giant public construction project to build an absolutely crucial addition to the port's infrastructure. In 1880, G. S. Greene, the Department of Docks' engineer-in-chief, presented a bold plan to redevelop the entire waterfront between Christopher and West 18th Streets, a distance of 4,430 feet, more than three-quarters of a mile. The plan included removing six city blocks: 13th Avenue and "all buildings, piers, earth and mud, west of this 250-foot street [West Street] to the depth of 25 feet below mean low water mark."[23]

Greene estimated the total cost—including purchasing the property; excavating 1,860,000 cubic yards of fill; and constructing twenty-one new piers—at $7.5 million. He expected the work to be completed in about eight years, but instead, twenty years of negotiations and court proceedings ensued. The land and buildings on just the six blocks between West 11th and Gansevoort Streets cost slightly over $5.9 million. This sum represented 79 percent of the original projected cost of the entire proposed project, an estimate that had included both the cost of land and construction of twenty-one piers.[24] Real estate development in the City of New York has always been contentious, as the Department of Docks learned in the 1880s and as it remains today.

Map 7.3 illustrates the redevelopment of the waterfront between West 11th and Gansevoort Streets in Greenwich Village, superimposed on the six blocks that were removed. One street (13th Avenue) disap-

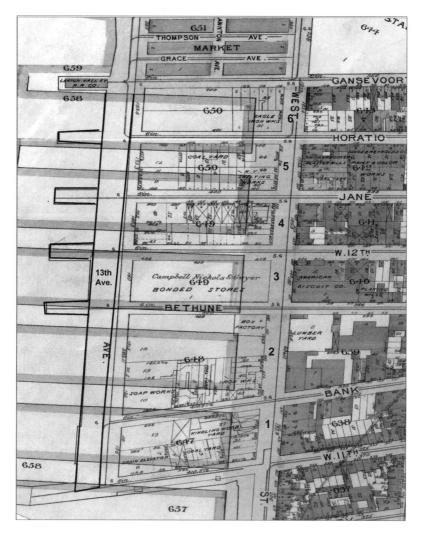

Map 7.3. Hudson River water-lots and the Department of Docks' pier, bulkhead, and made-land repurchase in Greenwich Village, from West 11th Street to Gansevoort Street. The Department of Docks repurchased six city blocks between 13th Avenue and West Street; cleared and excavated the land; let the Hudson River flood the area; and then built new, longer piers from West Street out into the river, eliminating 13th Avenue. *Source:* Created by Kurt Schlichting. Source GIS layer: historic street and pier maps, New York Public Library, Map Warper.

peared, as did the made-land between it and West Street. To acquire the needed made-land, the department repeatedly had to go to court to force the private owners to sell their land and buildings. Part of the extraordinary increase in cost resulted from the department's decision to once again begin the demolition of existing buildings and then excavate before the courts determined the final cost. The Department of Docks argued that if it waited for the final judgments, the project would never have been completed in a reasonable period of time.

In the end, Greene's plan resulted in five new piers, which the Department of Docks leased to the major transatlantic steamship companies for steep fees. The British companies had no choice, since the new Greenwich Village docks provided the only piers on the entire Manhattan waterfront that could accommodate their large passenger ships. Moreover, all costs for maintenance and dredging remained the responsibility of the steamship companies, who negotiated leases for the docks. To stay competitive, all the British steamship companies needed to remain on the Manhattan waterfront. Talk of moving across the Hudson River to the waterfront in Jersey City or Hoboken remained just talk. Even though officials in both of these cities offered incentives, the companies knew that their passengers did not want to end a luxurious Atlantic crossing on the other side of the river, a ferryboat ride away from their ultimate destination: Manhattan.

All of the major transatlantic lines anticipated a prosperous future carrying passengers and high-value freight across the Atlantic Ocean to New York. With long-term leases, the companies ensured their access to the waterfront.[25] For New York City, the lucrative leases for the new piers testified to the value of the rebuilt waterfront and the modernized maritime infrastructure built by the Department of Docks. The department's annual report for 1905 triumphantly pointed out that the new leases for the Greenwich Village piers would not only pay back the cost of construction, but also the bonds, as well as generate a surplus.

≈≈ Chelsea Piers

When the Department of Docks first proposed the Hudson River pier improvements, the plans included the Chelsea waterfront north of Gansevoort Street, from West 14th to West 30th Streets. Costs and delays postponed this work until after the completion of the new piers in Greenwich Village. Relentless technological innovations in ship design—in particular, the new, powerful steam turbines—and the fierce competition among the transatlantic lines led to the construction of larger and more luxurious ships. The planned *Titanic* and *Lusitania* measured over 800 feet long, and with still-bigger ships on the drawing boards, the new Chelsea piers needed to exceed these lengths. Even with the completion of the Greenwich Village piers, a number of the transatlantic shipping companies that had not obtained leases sent applications for new piers to the Department of Docks. In 1898, the president of the board, J. Sergeant Cram, argued for the addition of new piers in Chelsea. He contended that new piers would "cost $12,000,000, but not one cent would they cost the people of this city. We can rent them readily for sums that would realize 5 per cent on their cost."[26] Cram continued by emphasizing that with the city able to sell bonds at 3 percent, the new piers would pay for themselves and could be constructed in two years.

The proposed longer piers would extend beyond the established pier line in the Hudson River, and the Department of Docks again appealed to the Army Corps of Engineers to build 200 feet past that line. Once more, the appeal failed, and only one course of action remained. Just as with the new piers in Greenwich Village, the department would undertake the expensive and time-consuming process of repurchasing the inland water-lots and moving the bulkhead back toward the original shoreline. The department began the legal process of buying back the made-land between West 14th and West 23rd Streets owned by Clement C. Moore, the General Theological Seminary, and other grantees (see map 2.4). Given the difficulties and delays encountered earlier, the condemnation of these properties proceeded more quickly, but not inexpensively. The price for acquiring this land almost equaled Cram's optimistic estimate for the cost of the entire project. Between 1871 and 1920, the Department spent $42.6 million acquiring the bulkhead around the lower part of Manhattan Island and repurchasing water-lots. The Chelsea pier acquisitions alone accounted for a quarter of the total expenditure.

Construction of the new piers began in 1895 and continued for the next decade, completely transforming the Chelsea waterfront (map 7.4). At 16th Street, the new bulkhead stood 560 feet inland from the old bulkhead along 13th Avenue, an avenue that disappeared. To the north,

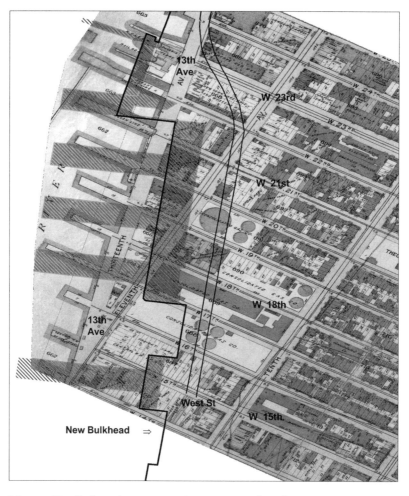

Map 7.4. The Chelsea piers were another massive and costly project. Once again, the Department of Docks purchased entire city blocks, excavated, and moved the new bulkhead back toward the original Manhattan shoreline. West Street was also relocated, and 1,000-foot piers were built out into the Hudson River. *Source:* Created by Kurt Schlichting. Source GIS layers: historic street and pier maps, New York Public Library, Map Warper; Chelsea piers map, New York City Planning Department.

at West 21st Street, the project extended the bulkhead past 13th Avenue and into the Hudson River. A new, wide extension of West Street in front of the new piers created open space that heightened the impact of the new piers' architecture.

Figure 7.3. The Chelsea piers, circa 1912. These new piers were the longest in the port and could accommodate luxurious ocean liners over 1,000 feet in length. *Source*: Shorpy Inc. / Vintagraph / Juniper Gallery.

Since the new piers served the luxury transatlantic passenger service, the Department of Docks engaged Whitney Warren, the architect for the new Grand Central Terminal, to design an elaborate façade along West Street. Warren employed the Beaux-Arts style to introduce a sense of grandeur. Previously, a typical pier shed included a utilitarian façade, adorned with garish lettering advertising the name of the railroad or shipping company. Warren's ornate portico faced the widened new section of West Street, embellishing the pink granite façade with ornamental figures decorating the cornice (fig. 7.3).

As work progressed, the *New York Times* hailed the new piers as "one of the most remarkable water-fronts in the history of municipal improvements . . . converting an extensive section of Manhattan's Hudson River waterfront into the most colossal steamship terminal." The *Times* praised the new piers' magnificent architecture, "so imposing that the foreign visitor will no longer find false impressions of the great city—no tawdriness in the way of piers."[27] The Chelsea piers created a new streetscape, bringing the City Beautiful Movement to New York's waterfront. Warren's Grand Central Terminal (1913), Carrere and Hastings's New York Public Library (1907), the Metropolitan Museum of Art's 5th Avenue façade (1902), and Pennsylvania Station (1910) all celebrated the rise of New York to a world-class city. Now transatlantic passengers would use the new Chelsea piers as the city's magnificent gateway to the Atlantic World. At Warren's Grand Central Station, the main concourse provided an even more-elaborate gateway for passengers arriving by train.

On February 21, 1910, the Chelsea piers officially opened when the White Star Line's *Oceanic* berthed at Pier 60. The major transatlantic passenger lines now moved to these nine new piers.[28] To cover the cost of the new piers ($4,505,540) and the purchase of the relevant water-lots, the Department of Docks negotiated long-term leases for a ten-year period, renewable for two more ten-year periods, extending from 1910 to 1940.[29] The steamship lines all believed the future of transatlantic passenger service to be boundless, and they planned to continue using the piers for the next several decades. Another challenge for the port arose when these steamship companies announced plans for even larger ships, with the White Star Line's new *Oceanic* to be over 1,000 feet in length. Once again, the department would have to build new, longer piers farther up the Hudson River, at Midtown.

With the Department of Docks' rebuilding of the Manhattan waterfront, both shipping and railroad freight traffic to the port continued to grow, especially along the Irish waterfront in Greenwich Village and Chelsea. In the decade and a half before World War I, the number of arriving and departing ships dealing in international commerce increased from 8,178 in 1907 to 10,580 in 1916.[30] A half century earlier, in 1860, the number of ships handling foreign commerce had totaled 3,960.[31] At the turn of the twentieth century, 50 percent of the total exports and imports for the entire United States moved through the Port of New York.

The number of ships in international service consisted of only a modest share of the vessels arriving and departing in the port. Yearly, thousands of coastal ships and smaller boats arrived via the port's waterway empires. In addition to shipping, the volume of freight handled in the port by the railroads reached a staggering 76 million tons in 1914.[32] With 20 million tons of this total inbound for Manhattan, New York Central freight cars carried almost 3 million tons directly onto Manhattan Island. The New Jersey–based railroads handled the remaining 17 million tons in their rail yards on the New Jersey side of the harbor and then floated the freight across the Hudson River to the Manhattan waterfront. Once on the New York side, the freight had to be landed at the freight terminals and then either hauled by drays and trucks to final destinations in Manhattan or carted to a pier, where longshoremen loaded a waiting ship.

≈ A Metropolitan Port and a Metropolitan Region

Even after the grand opening of the Chelsea piers, a series of systemic problems remained. Crowding along the shoreline did not lessen, and corruption, crime, and labor unrest persisted on the docks and in the surrounding neighborhoods where the longshoremen lived. By 1900, Manhattan no longer handled all of the ships and cargo arriving in the Port of New York. The Brooklyn and Staten Island waterfronts expanded. Across the Hudson, new piers were built in Jersey City and Hoboken, and some shipping companies relocated to the Jersey shore. That state's politicians and business leaders were more than delighted for the railroads to handle more freight on the New Jersey side of the harbor. The railroads thus saved the added cost of moving freight the "final mile" across the Hudson River to ships docked in Manhattan.

By the dawn of the twentieth century, the Port of New York included more than just the waterfront on Manhattan Island. Piers lined the Queens shore along the East River, as well as the shores of Brooklyn, Staten Island, Newark Bay, and the New Jersey waterfront cities of Bayonne, Jersey City, Hoboken, and Weehawken. A sprawling bi-state region evolved, encompassing numerous local governments. It included the largest city in the country, New York, and one of the smallest, Hoboken's 1 square mile. The multiple local governments each pursued their own agendas and sought to maximize their gains.

The Department of Docks viewed the cities across the Hudson River as rivals for the port's maritime business. The eight railroads in New Jersey vigorously complained about the cost advantage the New York Central enjoyed, with its direct rail service onto Manhattan Island. The

US Interstate Commerce Commission, however, continued to prohibit the New Jersey–based railroads from adding additional freight costs for hauling goods across the Hudson. Thus a manufacturing company in Manhattan paid the same rate to have freight delivered by the New Jersey railroads as by the New York Central.

In additional to the fight over freight rates, New York State and New Jersey clashed over plans to build longer piers out into the Hudson River, and both states had conflicts with the Corps of Engineers, which was concerned about the Hudson remaining navigable for large ships. New Jersey politicians, from the governor to mayors, resented the dominance of New York City in the Port of New York. No one political entity or interest group, however, exercised administrative oversight or coordinated development for the entire bi-state port.

William J. Wilgus, the New York Central Railroad's chief engineer, who planned the massive Grand Central project, realized that the transportation challenges in the port involved more than the railroads. The fragmentation of political and regulatory power in the region stymied all efforts to improve the maritime and railroad infrastructure throughout the port, in order to accommodate the ever-increasing number of ships and volume of freight. In 1909, he proposed the creation of an interstate metropolitan district, which would include all of the cities, towns, and villages in the bi-state region surrounding the Port of New York.[33] This district would form a quasi-governmental metropolitan agency, with delegated power from both New York State and New Jersey, to plan necessary improvements and exercise regulatory control over the entire port. Transportation experts in both railroads and shipping would lead the needed rational process to maintain the port as the busiest and most important in the country.

Almost all of the major interests in the port opposed Wilgus's plan. Each viewed any effort to change the status quo as a threat to its own parochial benefits. All of the railroads in New Jersey operated their own tugboats, lighters, and car floats, at great expense. New York City's Department of Docks worried that any efforts to coordinate with New Jersey would result in a loss of revenue from the wharves and piers in the city. The mayor and the City Council—and, behind the scenes, Tammany Hall, the longshoremen's union, and the mobsters who now controlled the piers—all saw the proposed district as a threat to patronage-based waterfront jobs and the spoils from graft and corruption on the docks. While Wilgus's plan received some support in both the New York State and the New Jersey legislatures, the strong political and business opposition to it could not be overcome.[34]

World War I created enormous pressure, leading to massive congestion, thus stymieing the movement of freight through the Port of New York. Soon after the United States entered the war on April 6, 1917, planning got underway to transport and supply the American Expeditionary Force (AEF). The AEF would ship out from New York, and supplies would follow. Troop strength was projected to reach 2 million by December 1918, and that number of soldiers would need 52,000 tons of supplies each day, all to be shipped across the Atlantic Ocean.[35]

By 1917, the demands on the country's railroads and the Port of New York in supplying the troops led to an almost complete breakdown of the transportation system. The inefficient means of moving freight across the Hudson River to ships bound for Europe could not meet the demand, and thousands of tons of freight piled up on the piers throughout the harbor. With the port paralyzed, trains bound for New York backed up all the way to Chicago, creating a serious shortage of freight cars. In December 1917, President Woodrow Wilson nationalized the railroads and appointed William McAdoo, a talented railroad executive who completed the Hudson & Manhattan Railroad passenger tunnels under the Hudson (today's PATH rapid transit system tunnels), as head of the US Railroad Administration. With his direct control over all of the railroads and shipping companies serving the Port of New York, McAdoo soon solved the freight crisis. His efforts treated the entire bistate port as one integrated system, which was administered by one set of executives, not by eleven separate railroads in New Jersey and New York, the Department of Docks, or the myriad shipping firms.

~~ The Port of New York Authority

In the war's aftermath, the idea for a metropolitan port district resurfaced. The success of McAdoo and the US Railroad Administration encouraged Wilgus and others to return to arguing for a metropolitan agency. The Russell Sage Foundation provided support for a regional planning model and, in 1922, established the Committee on the Regional Plan of New York and Its Environs. After seven years of research, the committee published its plan. It also established the Regional Plan Association, an organization to support regional planning.[36]

On the political front, in 1919, New York State and New Jersey, responding to pressure from the New York Chamber of Commerce and others, passed legislation to form the joint New York New Jersey Port and Harbor Development Commission. The commission's charge included undertaking a detailed study of the then current transportation problems in the port and making recommendations for change. One basic premise underpinned the study: McAdoo had succeeded in the earlier freight crisis because he had the power to regulate transportation in the entire port and region. Any solution to the intertwined transportation challenges had to embrace not just the Manhattan waterfront, but a bi-state metropolitan region that extended for a 50-mile radius around Times Square. By the 1920 US census, this region was home to over 8.1 million people.

A metropolitan region—crossing the long-established political boundaries of state, city, and local governments—represented a new conceptualization of the port, the city of New York, and the surrounding counties in New York State and New Jersey. In the larger region, New York City and Manhattan Island continued to play a leading role, but the continued prosperity and commercial success of the port depended on a complicated relationship with the surrounding areas. Each depended on the other. When the commission published its detailed *Joint Report with Comprehensive Plans and Recommendations* in 1920, the key recommendation called for the formation of a bi-state authority to manage both the Port of New York and all rail and shipping, in order to benefit the entire region.[37] In 1921, New York State and New Jersey passed legislation creating a new political entity: an "authority," the first with that title in the United States. The two states, in establishing the Port of New York Authority, anticipated that professional management, a rational planning process, and "better coordination of the terminal, transportation and other facilities of commerce in, about and through the Port of New York, will result in greater economies, benefiting the nation, as well as the states of New York and New Jersey."[38]

The new authority had a defined public purpose: to improve transportation in the Port of New York area. To accomplish this public good, the two states delegated to this new entity power to do what an elected local government usually did: borrow money and build public infrastructure. What set the authority apart was that the commissioners who ran the organization were appointed by the two governors, rather than being elected or chosen by local elected officials. The creation of the Port Authority was not without its issues, however. What oversight would the mayor of New York or the mayor of Jersey City or any other local elected official exercise? If the new Port Authority decided to relocate some of the shipping and freight handling from the Manhattan shorefront to the other side of the Hudson River, did it have that power? And if the Port Authority did, what would be the consequences? Serious questions regarding the power of independent public authorities—a mixture of public/private power—remain to this day.

Within a few years, the Port Authority's vision expanded to encompass all forms of transportation in the metropolitan region, especially the new revolution created by the advent of automobiles, trucks, and airplanes. Jameson Doig's masterful study tracks the expansion of the Authority's reach, creating an "empire on the Hudson" far beyond the intent of the original legislation.[39] The Port Authority gained control over the Holland Tunnel, the first vehicular tunnel under the Hudson River, in 1930. Traffic through the new tunnel exceeded projections, beginning from the first day it opened, and generated more than enough revenue to pay the construction bonds. In turn, the Authority used the surplus to fund other bridge and tunnel projects. The Port Authority also took over Newark Airport in New Jersey, plus LaGuardia and Idlewild Airports in Queens. When the Authority negotiated to take over Newark's municipally owned airport, the final agreement included that city's dilapidated piers on Newark Bay.

Beginning in 1927, the Port Authority expanded the transportation infrastructure throughout the metropolitan region. Conspicuous by their absence, however, are major improvements in rail facilities and the maritime infrastructure.[40] During the Great Depression, the decline in shipping led to fiscal difficulties for the Department of Docks. The Port Authority offered to buy all of the piers in Manhattan and Brooklyn and invest in modernization. A political firestorm erupted, so the Port Authority turned away from that proposal and increased its efforts to expand the automobile and truck infrastructure. In the next decades, the Authority built bridges, tunnels, and trucking terminals and expanded the region's four airports.

The Port Authority would not return its attention to the region's maritime infrastructure until 1962, and then it focused all of its energy and investment in Newark Bay and the waterfront in Newark and Elizabeth, New Jersey. With the container revolution underway, the Port Authority orchestrated the shift of the Port of New York from the Manhattan and Brooklyn waterfronts across the harbor to Newark Bay. The Port Authority's infrastructure projects from 1927 to 1973 included:

- Holland Tunnel, 1927 (the Port Authority took over operations in 1930)
- Goethals Bridge, 1928
- Outerbridge Crossing, 1928
- George Washington Bridge, 1931
- Bayonne Bridge, 1931
- Inland Terminal No. 1, 1932
- Lincoln Tunnel, south tube, 1937
- Grain Terminal and Columbia Street Pier, Brooklyn, 1944
- Lincoln Tunnel, second tube, 1945
- Newark Airport and Port Newark, 1947 (acquired under lease from the city of Newark)
- LaGuardia and Idlewild (now Kennedy) Airports, 1947 (acquired under lease from the city of New York)
- Teterboro Airport, 1948
- New York Truck Terminal, 1949
- Newark Truck Terminal, 1950
- Lincoln Tunnel, third tube, 1957
- Elizabeth, New Jersey, Port Authority Marine Terminal, 1962
- World Trade Center, 1973

This exodus of the maritime world to New Jersey would lead to cataclysmic change along the Manhattan waterfront.

[8]

Crime, Corruption, and the Death of the Manhattan Waterfront

I n the 1920s, the Port of New York continued to grow. No one then could have anticipated the turmoil soon to engulf the port, city, state, and entire country. New York City and the metropolitan region prospered during the Roaring Twenties, and the stock market reached an all-time high, creating thousands of Wall Street millionaires. Imports and exports through the Port of New York averaged $3.3 billion a year between 1920 and 1925 and, in the next five years, continued to increase to $3.7 billion a year.[1] Shipping filled the port, and revenue for the renamed Department of Docks and Ferries exceeded expenses throughout the decade, which ensured that the department paid the interest and principal on the dock bonds used to rebuild the Manhattan waterfront.

While the port flourished and the new Port Authority turned its attention to tunnels, bridges, and airports, a set of intertwined problems plagued the waterfront, now centered on the West Side of Manhattan. New York's continued industrial growth added to the ever-increasing demand for space on the waterfront. Railroad freight traffic to the city and port grew exponentially, and the New Jersey–based railroads' freight operations needed expansive rail yards, adequate warehouse space, and depot space on the Manhattan shoreline. They had to deliver freight to the Manhattan waterfront or lose traffic to the New York Central.

Across the Hudson River, the railroads acquired almost the entire waterfront of Jersey City and Hoboken, building large train yards and freight warehouses. They also constructed piers to move freight cars

Figure 8.1. The New Jersey shoreline, showing the railroad yard, piers, and freight sheds. All of the major railroads in the country had to be able to bring freight to the Port of New York. The New York Central was the only railroad with direct access to Manhattan Island, with two freight yards along the Hudson River waterfront. The tracks of seven other railroads serving the port and the city ended at the New Jersey waterfront, a mile from Manhattan. *Source*: New York Central System Historical Society

onto barges and then be towed across the Hudson (fig. 8.1). After the Department of Docks began to rebuild the New York waterfront, the railroads secured long-term leases for freight depots on the piers along the West Side of Manhattan.

～ The West Side of Manhattan

To move freight from the New Jersey rail yards to Manhattan required each of the railroads to operate a small fleet of tugboats, lighters, barges, and car floats (fig. 8.2), a costly operation. The Pennsylvania Railroad's fleet included 55 tugboats, 226 barges, 124 car floats, and 9 lighters with steam cranes.[2] All of the other major railroads serving the port—the Erie; Lehigh Valley; Central of New Jersey; Philadelphia & Reading; West

Figure 8.2. A car float leaving Jersey City for New York City. To deliver freight the final mile to the Manhattan waterfront required an expensive trip. One means of doing so was to load railroad freight cars onto car floats and tow them across the Hudson River. On the Manhattan side of the river, the contents could either be moved to one of the numerous freight depots adjacent to the piers or unloaded directly from freight cars on the car floats, which were tied to the bulkheads on West Street or South Street. *Source*: New York Central System Historical Society

Shore; and Delaware, Lackawanna & Western—also operated their own tugboats and marine equipment, adding more expense.

To compound the price disadvantage, the US Interstate Commerce Commission (ICC), which had regulated the railroads since 1877, set the same rate for a ton of freight delivered to Manhattan or to New Jersey. Despite repeated appeals from the railroads and New Jersey interests, the ICC maintained a uniform freight rate throughout the Port of New York, giving the New York Central a decided advantage. The Central could bring freight by rail directly onto Manhattan Island, while the New Jersey–based railroads faced a significantly higher cost to deliver a ton of freight the last mile across the Hudson River (see map 6.1).

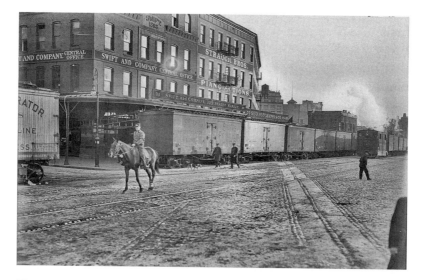

Figure 8.3. New York Central Railroad train and freight cars on 11th Avenue. A New York Central employee is shown riding a horse in front of the oncoming train, to warn of danger. The refrigerated freight cars were delivering dressed beef from Chicago to Swift & Company, located on 11th Avenue. The railroad's charter granted it the right to have its tracks on New York City streets, a right the company defended vigorously, despite the danger its trains—running day and night—posed. *Source*: Digital collections, New York Public Library

On Manhattan's West Side, the New York Central maintained its priceless monopoly on direct rail access to the island. The railroad's tracks and trains ran down the middle of the city's streets on 11th and 10th Avenues, below 30th Street, putting pedestrians at great risk (fig. 8.3). Newspapers, residents, local businesses, and politicians complained loudly about the Central's use of city streets, referred to as the West Side problem. The Central vigorously defended its rights, based on the Hudson River Railroad's 1845 charter. Ira Place, the New York Central's chief legal counsel, testifying before a state commission set up to investigate the West Side problem in 1917, emphatically stated: "We got certain rights from the State. The State gave us the right to build where we did build under the condition that we got the consent of and assent of the city and when we got it . . . that constituted our rights."[3] The commission concluded that the Central's trains on 10th and 11th Avenues created "a steady growing menace to life and limb and property and Tenth Avenue became notorious as 'Death Avenue.'"[4]

For the next twenty years, the Central tenaciously defended its legal rights to run trains on the West Side streets, a crucial advantage over their rivals, who were a mile across the Hudson in New Jersey.

New York City's ongoing battle with the Central represented another clash between public and private interests. In 1845, the State of New York had granted the Hudson River Railroad the absolute right to operate on the city of New York's public streets. At that time, the West Side of Manhattan Island and its waterfront remained undeveloped, and the East River shoreline served as the heart of the Port of New York. As both the city and the port grew, the maritime world moved to the Hudson shore. Pressure arose to have the city and the state remove the tracks and trains and restore public control over the streets, but the West Side problem was not resolved until the building of the High Line in 1928.

The New Jersey railroads leased piers and bulkheads along the Hudson River for freight depots. Loaded car floats were towed across the river and tied to the bulkheads, in order to unload at the shorefront freight depots. With space at a premium and freight volumes increasing, a number of the railroads purchased land across West Street and built inland depots (fig. 8.4). Small switching engines pulled the freight

Figure 8.4. Streets in the West 20s, showing some of the Hudson River piers, train tracks, and inland freight depots. Ships, tugboats, lighters, and ferries crowd the docks along the Hudson River. Across West Street, the New Jersey–based railroads had their own freight yards and freight depots. The railroads, shipping companies, and ferries competed for limited space on the Hudson River waterfront. *Source*: TrainWeb, courtesy of Philip Goldstein

Figure 8.5. West Side congestion, showing piers (*left*) and a moving train (*upper right*). The competition for space on the waterfront created gridlock on West Street. Horse-drawn drays, cars, and trucks crowd 11th Avenue. To the north, a New York Central freight train has brought all traffic to a halt. *Source*: Digital collections, New York Public Library

cars off the car floats and across the street to be unloaded, adding to the congestion on the West Side.

The sheer volume of freight created daily gridlock on West Street and the adjacent streets and avenues (fig. 8.5). Thousands of horse-drawn drays and the new automotive trucks hauled freight a second "last mile" to destinations in Manhattan, and then picked up freight to be delivered back to the inland depots or directly to the piers. Chaos ensued, and everyone complained about delays, lost shipments, damaged goods, and stolen property.

By 1914, the number of freight cars on the waterfront and city streets totaled almost 1 million a year. The Central alone hauled 270,998 freight cars back and forth between the 60th Street rail yard and St. John's Park Depot, north of the Battery—all on the city streets. Two of them, 10th and 11th Avenues, served as another freight yard for the New York

Table 8.1.
Tonnage and number of freight cars from New Jersey to Manhattan, 1914

New Jersey rail lines	Tonnage (lbs.)	Freight cars
West Shore	451,940	81,888
Delaware, Lackawanna & Western	476,943	77,186
Erie	663,666	116,974
Pennsylvania	1,478,004	226,312
Lehigh Valley	342,313	47,588
Central of New Jersey	607,413	97,112
Baltimore & Ohio	406,266	50,714

Source: New York New Jersey Port and Harbor Development Commission, *Joint Report, with Comprehensive Plans and Recommendations* (Albany: J. B. Lyon, 1920).

Central. In addition, the New Jersey–based railroads hauled 4.4 million tons of freight across the Hudson River at great cost, which they could not pass on to their customers (table 8.1).[5]

～～ William Wilgus and the West Side Problem

In 1894, the State of New York established the Rapid Transit Commission to regulate the horse-drawn street railways and elevated railroads, as well as to oversee planning for the city's first subway, the IRT in Manhattan. The commission also had to deal with the Central's tracks on the West Side. Political pressure increased, and in 1906 the state passed the Saxe Law, which directed the Rapid Transit Commission to remove the Central's tracks from the public streets on Manhattan's West Side. The Central continued to vigorously argue that it had the legal right to use the city streets and took the issue to court. The New York Legislature, frustrated by the intractability of the railroad, replaced the Rapid Transit Commission with the Public Service Commission (PSC) and directed the PSC to find a solution for the problems along the Hudson waterfront. The PSC hired William Wilgus—who by then had become the president of a consulting company, the Amsterdam Corporation—to study the West Side problem and recommend solutions.

From the outset, Wilgus recognized that the problems along the waterfront involved more than the Central's tracks. The seven New Jersey–

based railroads occupied more space than all the shipping companies combined: 28.9 percent of the 2.5 miles between the Battery and Gansevoort Street.[6] On West Street, the railroads had constructed big freight depots on the bulkhead, where thousands of drays and trucks came each day to pick up and deliver freight, creating perpetual gridlock.

Wilgus proposed a bold solution: to construct a freight subway under the streets of Lower Manhattan, connected via tunnels under the Hudson River to a massive train yard in New Jersey's Meadowlands, to be shared by all the New Jersey railroads (fig. 8.6).[7] In the Meadowlands, the railroads would load freight onto subway cars, to be deliv-

Figure 8.6. (*Below and opposite*) William Wilgus's proposed small-car freight subway. Wilgus's plan would have eliminated the need for the New Jersey–based railroads to float freight cars across the Hudson River. *Source*: Redrawn by Bill Nelson based on original documents from the William J. Wilgus Papers, Manuscripts Division, New York Public Library

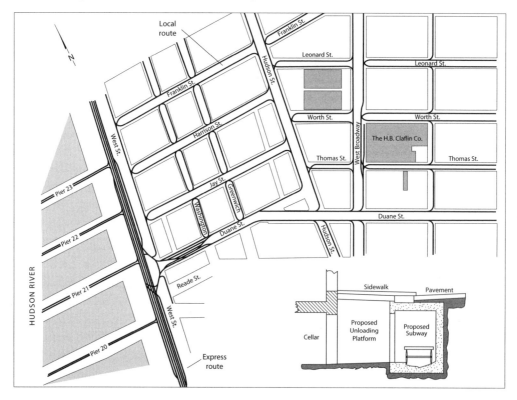

ered underground to eleven inland freight depots in Lower Manhattan. For outgoing freight, the process would be reversed. Wilgus's plan would eliminate the need to float freight across the Hudson. Instead, it would be delivered via tunnels under the river.

Wilgus estimated the then current cost of moving freight across the Hudson at $33 million a year. In his report, he pointed out that the cost to move a ton of freight from New Jersey to Manhattan exceeded the cost of hauling the same quantity of freight from Boston to New Jersey, a distance of 200 miles. For the "last mile" across the Hudson, a cumbersome, inefficient system generated enormous costs for the rail-

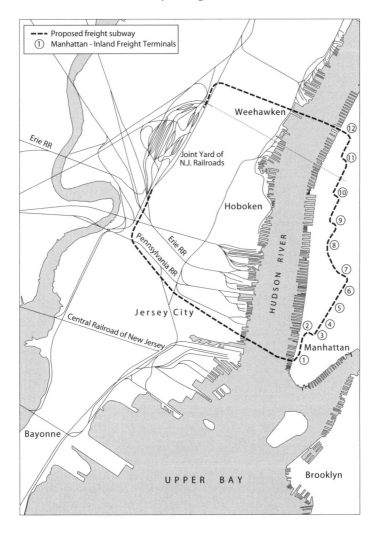

roads, which they could not pass on. A freight subway also eliminated the need for the New York Central to keep its rail tracks on city streets. Since the New Jersey railroads would be able to deliver freight directly to inland depots in Manhattan, the competition for pier and bulkhead space along the Hudson River would diminish. In turn, more pier space would open up for the ever-increasing number of ships using the port.

The Wilgus plan generated support from fellow engineers, and it had favorable coverage in the press.[8] From the railroads, Wilgus encountered nothing but opposition. The fiercely competitive railroad lines remained suspicious of any plan based on cooperative efforts, even if it promised significant cost savings. The Central, moreover, saw no reason to give up its monopoly on the West Side. Eight years later, the newly established Port Authority's first plans to solve the railroad problem in the port included a freight subway under the Hudson River, with several stations for loading and unloading freight in Manhattan.[9] Once again, the railroads' opposition killed a second plan for a freight subway under the Hudson, which further pushed the Port Authority to turn to bridges and highways as a solution for moving freight in the metropolitan region. A decade later, in 1928, the New York Central, the city, and the state finally reached an agreement to share the cost of an elevated railroad down the West Side of Manhattan, removing the tracks from the streets (see fig. 9.4). While the High Line arrived too late to prevent the decline of the waterfront, against all odds the structure survived, to be reborn in recent years as a public park.

∼∼ Storm Clouds

On Black Thursday—October 24, 1929—the stock market lost 11 percent of its value. Despite intervention by the larger banks, the market continued to decline, and by July it had lost 90 percent of its value. Soon banks began to fail, and manufacturers went out of business. Unemployment reached epic proportions, and hunger stalked the country. In New York, the unemployed begged for food, and thousands of families were evicted from their apartments. With the economy in free-fall, imports and exports plummeted to an average of $1.4 billion a year in the 1930s, 62 percent less than in 1928. On the piers, far fewer ships arrived or departed as the economic situation deteriorated world-wide. Thousands of longshoremen found no work. Despair and hunger haunted the Irish waterfront.

The rising demands of World War II ended the Great Depression, and the economy in the metropolitan region recovered as the United States became, in the words of President Franklin Roosevelt, the "arsenal of democracy." The Port of New York served as the major point of embarkation for troops bound for Europe, and millions of tons of supplies followed. In 1943, the railroads serving the port hauled 139,000,000 tons of freight, 72 percent more than in 1940.[10] Careful planning avoided gridlock in the port, and the trains ran efficiently. For once the railroads cooperated.

World War II ended with a sense of triumph and ushered in a period of unparalleled growth and prosperity. At the end of the war, the United States emerged as the world's economic powerhouse. With the rebuilding of Europe under the Marshall Plan, the Port of New York flourished: shipping increased, and longshoremen again found employment. At this point, the city's Department of Docks and Ferries underwent another name change, to the Department of Marine and Aviation. The department announced a series of plans to modernize both the piers along the Hudson and East Rivers and those in Brooklyn and Staten Island, though without much success.

~~ The Port's Dark Underbelly

In the nineteenth century, criminals plagued the piers and docks of Manhattan Island. These dock rats brazenly stole freight under the very noses of the undermanned harbor police.[11] Unscrupulous boarding house owners fleeced the sailors. Bars sold cheap liquor to the longshoremen, who shuffled to the nearest saloon if not picked in the shape-up. Tammany Hall controlled all work on the docks, and only those loyal to the corrupt Democratic political machine found employment. The dockworkers made sure to give an appropriate "contribution" to their Tammany ward heelers. Even worse corruption followed in the twentieth century.

Irish mobsters muscled their way to control of the West Side docks prior to World War I. Owen "Owney the Killer" Madden and Tanner Smith ruled over the lucrative stevedoring contracts on a number of piers. Stevedoring companies managed the handling of cargo and hired longshoremen at the daily shape-up.[12] For the longshoremen, the tyranny of the shape-up persisted. A US Senate investigation of labor issues on the docks in 1912 interviewed John F. Riley, a former stevedore who had become an organizer for the International Longshoremen's Association (ILA). Riley described in detail how the shape-up worked on the Chelsea docks. When asked how many men worked on a typical day, he responded, "about 2,500." When asked how many men hung around for that work, he replied, "5,000."[13] Riley explained that the individual longshoremen were picked with a tap on the shoulder. If not chosen, they often drifted to the nearest saloon. This day-labor system led to the longshoremen's reputation for drinking: "That is the whole cause, in my opinion, the men being idle every day, and a ship is laid out there, and the men don't get to work and have to lay around there and wait for one, and a longshoreman has to be ready at all times, ready to take a call. He may he called at 9 o'clock or he may be called at 9:30 or 10 or up to 11 . . . and it keeps a man laying around there, and if he is a drinking man he is liable to go to a saloon, and that has caused the down grade of our men, having to lay around and frequent the saloons, and that is the only place he can frequent, and that is making the home life and conditions bad for him."[14]

Riley estimated the earnings of men working the Chelsea docks at about $8 or $10 a week for three or four days' work, if they were picked in the shape-up. To work those three or four days, the men had to be available on the waterfront six or seven days a week. A dockworker's total earnings averaged $500 a year. Pier rules required the men to leave the pier for lunch or dinner, and most piers did not have toilets or a first aid station. A longshoreman suffering a serious injury—one requiring months of rehabilitation—while on a job would receive $100. If he were to be out for only three or four weeks, "the chances are he would get nothing—9 out of 10 times."[15]

Reformers saw the shape-up as an evil system that exploited the longshoremen and proved to be the source of corruption on the waterfront:

> It is a sight to sadden the most callous, to see thousands of men struggling for only one day's hire; the scuffle being made the fiercer by the knowledge that hundreds out of the number there assembled must be left to idle the day. . . . In the middle of the Depression here [the New York waterfront], when good men

were going hungry, begging for work, five hundred men lined up in a semicircle in front of the pier. . . . A new hiring boss took one look at their faces and lost his nerve. He threw the hundred brass checks in the air. Five hundred men fought one another like animals for a hundred brass checks and a half day's work. That's the shape-up in New York, the greatest port in the world, and this is the twentieth century.[16]

The army of longshoremen and dockworkers filled the tenements in the Irish waterfront neighborhoods. With no guarantee of employment, they and their families lived precariously, on the edge of poverty.

≈≈ The ILA and "President for Life" Joe Ryan

The exploitation of longshoremen found a willing partner in the International Longshoremen's Association. The ILA, part of a national effort to organize waterfront labor, was founded on the Great Lakes in the 1870s. Its first New York local, Local 791, opened in 1908, with an office in Chelsea. Joseph Ryan, eventually the ILA's "president for life," joined the union in 1913. Ryan, a product of the Irish waterfront, had a sixth-grade education and worked briefly as a longshoreman, before an injury ended his career on the docks. He rose through the ranks and became president of the Atlantic Coast District of the ILA in 1927, which included all the locals along the Manhattan and Brooklyn waterfronts, including the "mother local," Local 791. Ryan lived in Chelsea and went daily to the 8 a.m. Mass at Guardian Angel Church. He remained a fast friend of its pastor, Monsignor John J. O'Donnell. Through O'Donnell, Ryan kept in contact with the Archdiocese of New York. At the annual ILA dinner in the Waldorf Astoria Hotel, O'Donnell and the archbishop sat at the head table with Ryan and his senior henchmen.

According to a New York City district attorney's investigation of Ryan and the ILA, Ryan had a history of "surrounding himself with other 'gangster associates' from former bootleggers such as Owney Lawlor . . . to various criminals (Johnny Dunn, Thomas Burke, and Victor Patterson, an 'ex-convict with a criminal record dating back to 1915') who'd all become officials in Ryan's union."[17] On January 8, 1948, Johnny "Cock-eye" Dunn murdered Andy Hintz in the hall of Hintz's apartment at 61 Grove Street in Greenwich Village. Hintz—the hiring boss for Pier 51 at

Jane Street, on the Hudson River—had refused to let Dunn muscle in on his pier. Hintz's murder was just one among hundreds that occurred on the waterfront, but no one ever talked. When asked why he hired criminals such as Dunn, Ryan responded: "We welcome ex-convicts. . . . We believe every man, regardless of what he has done, has a right to a second chance."[18]

In Nathan Ward's book about waterfront crime and corruption, *Dark Harbor*, Joseph Ryan and the ILA play a prominent role.[19] Waterfront corruption and the domination of the longshoremen seemed intractable. An unscrupulous union, criminals, and mobsters had the waterfront in an iron grip, which condemned the longshoremen and their families to a life of misery. Malcolm Johnson, a reporter for the *New York Sun*, condemned Ryan and the ILA: "The International Longshoremen's Association as now constituted is a model for dictatorship. . . . The Ryan rule does not promote the welfare of the union members, and . . . many acts of the union leadership are, in reality, antiunion and antilabor. . . . The longshoremen are unable to help themselves measurably because of their union's laws and [the] practices of the entrenched leadership [are] concerned only with maintaining its power."[20]

≈ On the Waterfront: The Movie and the Reality

On the night of July 28, 1954, at the Astor Theatre on Broadway at West 45th Street, the Hollywood movie *On the Waterfront* opened. The film, starring Marlon Brando, Karl Malden, and Eva Marie Saint, received stellar reviews: "A small but obviously dedicated group of realists has forged artistry, anger, and some horrible truths into *On the Waterfront*, as violent and indelible [a] film record of man's inhumanity to man as [has] come to light this year. . . . An uncommonly powerful, exciting and imaginative use of the screen by gifted professionals. . . . Scenarist Budd Schulberg . . . director Elia Kazan . . . the principals headed by Marlon Brando . . . convincingly have illustrated the murder and mayhem of the waterfront's sleazy jungles."[21] The movie, nominated for twelve Academy Awards, won eight Oscars, including to Budd Schulberg for Best Screenplay, to Elia Kazan for Best Director, and to Marlon Brando for Best Actor.

Hailed for its authentic portrayal of the brutal lives of the longshoremen, the movie depicts the struggles of the Irish dockworkers and their families. The corrupt ILA and the mobsters terrorized the longshoremen and their families, using violence to demand payoffs in return for hiring a man for a day's work on the piers. *On the Waterfront* captures the gritty day-to-day life on the piers and in the waterfront neighborhoods.

Karl Malden played an Irish priest who ministered to the immigrant population on the waterfront; at that time, tough city priests had become stock characters in Hollywood movies. Malden's character, the crusading Father Barry, was not fictional. Fr. John (Pete) Corridan, SJ (1911–1984), worked at the Xavier Labor Institute at Xavier High School, which was in a Jesuit parish on West 16th Street, a few blocks east of the Chelsea piers. He attended Regis High School in Manhattan, entered the Jesuit order, and was assigned to Xavier in 1946.

Corridan and other young Jesuits—the labor priests—came of age during the 1930s and 1940s, when the Catholic Workers Movement, led by Dorothy Day, championed the rights of workers to form unions and collectively bargain for higher wages and better working conditions.[22] Day founded an influential newspaper, the *Catholic Worker*, in 1931. Inspired by Day and papal labor encyclicals, the labor priests saw their ministry as dedicated to the "human dignity of all people."[23] To follow in the footsteps of Dorothy Day, hewing to the spirit of the encyclicals and Catholic social teaching, meant to work for social justice for the longshoremen and their families, most of whom were Irish Catholics. Corridan and other labor priests ministered to more than their parishioners' spiritual needs; they dedicated themselves to bringing social justice to the Manhattan waterfront. Corridan joined the Xavier Institute in 1946 and, with its director, Fr. Philip Carey, SJ, taught evening classes to waterfront workers. The classes focused on labor organizing, collective bargaining, and the details of federal and state labor laws. Corridan also spent time on the waterfront and in the adjoining neighborhoods, talking to the workers and gaining insight into the exploitation and violence on the piers.

Corridan's deep knowledge of the corruption on the waterfront led to two crucial relationships: one with the crusading journalist Malcolm Johnson, and the other with the Hollywood screenwriter Budd Schulberg. Johnson, a staff writer for the *New York Sun*, covered organized crime and the labor rackets. In 1948, the *Sun* published a series of articles he wrote, entitled "Crime on the Waterfront." The first story ran in

October, and Johnson's reporting created a sensation. He described in detail the corruption of Joe Ryan and ILA, the control of the piers by mobsters, and the violence both groups employed. The Irish dockworkers maintained a code of silence, based on a strong sense of tribal loyalty. Behind the union and the collusion of the Democratic political machine, and abetted by the silence of the Catholic Church, stood the shadowy businessmen who ran the waterfront companies: shipping, stevedoring, construction, and transportation. William McCormack, "Mr. Big," amassed a fortune with his myriad waterfront business interests and maintained strong ties to the Archdiocese of New York. McCormack, a major contributor to the archdiocese, attended Mass each day and supported the waterfront parishes, including Guardian Angel in Chelsea, Joe Ryan's church.

Malcolm Johnson's reporting for the *Sun* won a Pulitzer Prize in 1949. For years, though, Johnson kept an important secret: his waterfront source, Fr. John Corridan, SJ. James Fisher's compelling saga, *On the Irish Waterfront*, identifies the crucial role Corridan played: "It is unlikely that [Malcolm] Johnson or any other reporter could have dissected the loading racket [and waterfront corruption in general] without the benefit of Corridan's empirical and analytical resources."[24]

Elia Kazan, the controversial director of *On the Waterfront*,[25] met Corridan in early 1952. Drawn to crafting a waterfront movie by Johnson's reporting, Kazan hired a noted screenwriter, Budd Schulberg, to create the script. Corridan became Schulberg's inside source and strongly influenced the final version of the movie's screenplay. Corridan and Schulberg walked the waterfront in Greenwich Village and Chelsea, where the priest introduced the scriptwriter to an array of waterfront characters. Schulberg later attributed the success of his screenplay to close collaboration with the Jesuit labor priest.

Kazan's film captured the viciousness of waterfront life. In the gritty shape-up scene, the longshoremen are cast as victims in a violent world. They find solace in the neighborhood bar and strength in their ethnic ties. A code of silence pervades the waterfront, and the longshoremen refuse to "name names" of the hoodlums who control the docks. The code of silence rests partly on the shared ethnic heritage of the Irish laborers and partly on the fear of violent retribution awaiting those who talked. The latter was depicted in the brutal beating of Terry

Malloy (Brando's character) in the movie and attested to by the real murder of Andy Hintz in 1948.

≈ The Crusade on the Waterfront

Fr. Corridan persisted in his efforts for social reform on the waterfront, focusing on the ILA and testifying before the New York State Crime Commission's investigation of union corruption. On January 12, 1953, the *New York Times* ran a story about Corridan and a report he wrote for the Crime Commission. In the report, the priest condemned the ILA as a "company union," dominated by "gangsters." He called for an end to the shape-up and advocated opening hiring halls, to be run by the New York State Employment Service, not the union.[26] He also urged the Port Authority to acquire the waterfront in New York Harbor, in order to remove the piers from the control of the Democratic machine, the corrupt union, and the mob. Corridan encountered unexpected opposition from rank-and-file dockworkers, who thanked the priest for his spiritual aid but added, "Please let us run our union."[27] This sentiment was echoed by Monsignor John O'Donnell, Joe Ryan's pastor at Guardian Angel Church, who openly challenged Corridan.

Corridan's efforts, the movie, Malcolm Johnson's exposé, the investigation by the State Crime Commission, and US Senate hearings—the latter chaired by Senator Estes Kefauver—all brought enormous pressure to bear on the ILA. The American Federation of Labor (AFL), led by George Meany, also entered the fray. In August 1953, he suspended the ILA for pervasive corruption. The AFL set up a rival waterfront union and called on the National Labor Relations Board (NLRB) to hold elections in the port, to determine which union would represent the longshoremen and dockworkers.

Corridan gave his unequivocal support to the AFL, sending open letters to the longshoremen to persuade them to throw out the ILA. He predicted that the AFL would win by a better than three-to-one margin, placing his faith in the dockworkers to choose good over evil.[28] The final outcome, however, was a defeat for the AFL and Corridan: 9,060 votes to 7,568. Charges of intimidation and voter fraud led the NLRB to hold a second election in spring 1954. Corridan again campaigned in this election, to no avail. The ILA won by an exceptionally small margin—9,110 to 8,791—but win it did, with the result certified by the NLRB.

For Corridan and the Jesuit labor priests, the election's outcome proved to be devastating. Corridan reportedly stated, "I've lost." Fisher concludes that "the failed 1954 electoral campaign had finished Pete Corridan on the waterfront of history."[29] Corridan not only did not sway the election, but the Archdiocese also lost its patience with the waterfront priest and pressured the Jesuits to remove him from the Xavier Labor Institute. His Jesuit superior transferred him to LeMoyne College in Syracuse, New York, his ministry forgotten on the Irish waterfront.

William McCormack, "Mr. Big," died in 1965. His Mass of Christian Burial, held at St. Patrick's Cathedral on 5th Avenue, was concelebrated by Cardinal Francis Spellman, Bishop Fulton J. Sheen, and Bishop Terence James Cooke. Fr. John (Pete) Corridan, SJ, died quietly on July 1, 1984, in the Jesuit residence at Fordham University's Rose Hill campus in the Bronx. Few attended the Mass and the following brief ceremony at the Jesuit cemetery nearby. Budd Schulberg went to Corridan's funeral and described the turnout as "pathetic." Schulberg later wrote: "But there will never be another Father John (Pete) Corridan. He brought the story of the Crucifixion from Calvary to the piers of the 'Pistol Local' [ILA Local 791–Chelsea]. And right now I can see him there, not resting in peace but giving hell to the Bill McCormacks, the Joe Ryans, and even some to the all too accommodating monsignors and cardinals who messed up what he considered the true obligation of the Church. *Vale atque ave* [hail and farewell], Father Pete."[30]

How can one explain the dockworkers' tribal loyalty to the corrupt ILA and their stoic acceptance of the brutal way of life on the waterfront? The noted sociologist Daniel Bell—who, before his academic career, wrote about labor for *Forbes* magazine—devoted a chapter in *The End of Ideology* to the "racket-ridden longshoremen."[31] In 1960, looking back over four decades, Bell asked, "How could such a corrupt and fetid state of affairs have continued for so long?" "Corrupt" and "fetid" aptly described the life of the longshoremen and their families. The waterfront neighborhoods remained homogenous and self-contained:

> There is the bar, the social club and occasionally the parish house as the center of community life. St. Veronica's Parish on the West

Side is a typical longshoremen slum. The men live in a narrow band of tenements. . . . Slashed across the center is Pig Alley . . . a junk-strewn thoroughfare. . . . At night the souses lurch along until they slide down into the street to sleep off their drunk. . . . On Sunday the little girls walk self-consciously in their organdies to communion. Most of the people live in the houses where they were born, and die there. . . . But withal, it is a community and no one is a "cop-hollerer."[32]

No one broke the code of silence, just as Brando's Terry Malloy had avowed in *On the Waterfront*: "I don't know nothin', I ain't seen nothin', I'm not sayin' nothin'."[33]

In the long run, Corridan's efforts were not in vain. The success of *On the Waterfront* and the long investigation by the State Crime Commission finally led to reform. In 1953, New York City's district attorney indicted Ryan on charges of stealing from the ILA, using the money to pay his membership dues at the Winged Foot Country Club and to purchase Cadillacs and luxury cruises. The AFL expelled the ILA for gangsterism. Ryan beat the charges for theft, but then the federal district attorney indicted him for bribery and tax evasion. The union finally fired him, but they awarded their "president for life" a $10,000-a-year pension. When Ryan died in 1963, his funeral at Guardian Angel Church included "a modest funeral cortege—seventeen limos, four flower cars— [that] drove past the docks."[34]

Governor Thomas Dewey of New York, who built his reputation fighting organized crime as the district attorney for New York County, and Governor Alfred Driscoll of New Jersey urged their legislatures to form another bi-state agency to regulate the waterfront labor system. In 1954, the two states established the Waterfront Commission of New York Harbor, to "impose a uniform hiring system, eliminate the shape-up and outlaw 'public loading' throughout the port."[35] All longshoremen would be registered and report each day to hiring halls, run by the Waterfront Commission, where work would be assigned by seniority. No longer would the gangsters control the piers.

The ILA and even the longshoremen bitterly opposed the Waterfront Commission. Lawsuits and wildcat strikes followed. To this day, opposition to the commission persists, and corruption and organized crime on New York's waterfront have not disappeared. The Waterfront

Commission's annual report for 2012–2013 candidly identifies the problems the agency still faces, ones that have persisted on the waterfront for hundreds of years:

> Change in culture will not come easily to an industry with a long and intractable history of corruption and racketeering. Under the previous administration, the hiring, training and promotion practices of the industry led to no-show and no-work jobs, favoritism and nepotism, the abusive and illogical interpretation of existing collective bargaining agreements, and the impact of such practices both on the competitiveness of the Port and on the morale and career prospects of decent, hard-working Port employees. Such no-show and no-work positions, generally characterized by outsized salaries of up to $460,000, are overwhelmingly given to white males connected to organized crime figures or union leadership.[36]

～ Decline of the Manhattan Waterfront

Despite reform efforts, Manhattan's dominance of the Port of New York slowly waned. If the waterfront's rise to a position of ascendancy by 1860 was quick, the decline, as Benjamin Chinitz describes, was "a century of slowing down."[37] In absolute terms, the volume of foreign trade passing through the port increased after World War II, but the port's share of the total amount of trade in the United States dropped. New York's portion of the country's exports shrank from 24 percent between 1923 and 1929 to 10 percent in 1955. Imports declined to 21 percent over the same time period.[38]

While the Chelsea waterfront project in the 1930s created state-of-the-art pier space, during the Depression and World War II, expenditures for pier upgrades or replacement fell to less than $150,000 a year.[39] The Department of Marine and Aviation competed with other city priorities for capital expenditures. With the automobile and truck revolution well underway, highways and bridges had gained precedence.

Port facilities in Baltimore, Philadelphia, and Norfolk, Virginia, were expanded during the war years. Barge and railroad freight traffic down the Mississippi River increased and brought more business to the ports of New Orleans and Mobile. The Port of New York handled 151,784,000 tons of cargo in 1951, while Boston, Philadelphia, Baltimore, Norfolk, Mobile, and New Orleans combined handled 353,163,000 tons.[40] To compound matters, the Port Authority began to invest in modern facilities along the shore of Newark Bay and eventually would build a huge, modern container port in Newark and Elizabeth, New Jersey.

Fewer and fewer ships used the piers in Manhattan and Brooklyn. For the Department of Marine and Aviation, the decline resulted in vacant piers, as shipping companies were unwilling to sign long-term leases. In August 1951, fourteen piers around Manhattan lay vacant—something unheard of in the 1920s—resulting in $1.1 million in lost revenue. The department's revenue in 1950 totaled $9.9 million, but, after paying debt service, it ran a deficit of over $2 million. Over the first sixty years of their operation, from 1870 to 1930, the city's piers had always generated a surplus. Vacant piers meant lost revenue, as well as fewer jobs for the longshoremen and other dockworkers who depended on the waterfront for their livelihood.[41]

≈≈ The End of Manufacturing

Accompanying the precipitous decline in Manhattan's shipping and cargo tonnage, the city's manufacturing economy similarly fell. New York had prospered during the era of American industrialization by combining low-cost freight delivery to the shore, factories nearby, and a surplus labor force from the city's high levels of immigration. On crowded Manhattan Island, manufacturing companies occupied vertical space in multistory factory buildings. Horse-drawn drays hauled freight over the short distances from factory to shoreline. Manhattan led the country in manufacturing after the Civil War, and the city accounted for over 50 percent of all manufacturing employment in the metropolitan region until the turn of the century. Manufacturing in Brooklyn, near the waterfront, also boomed.[42]

Waves of immigrants had found employment in the city's thousands of manufacturing firms, from garment sweatshops in the Lower East Side to the Novelty Iron Works near the East River piers. With the coming of trucks and automobiles, however, the factories and their workforce were no longer tied to the shorefront. They both would decentralize and move throughout the metropolitan region. In part, Manhattan hemorrhaged manufacturing jobs because firms relocated to facilities with more horizontal space, to accommodate modern continuous assembly lines. Between 1956 and 2013, Manhattan lost 357,827 such jobs, a drop of 95 percent, and Brooklyn lost 99 percent. The city entered a period of decline, a virtual death spiral, as factory jobs vanished—never to reappear—for tens of thousands of residents. With this loss of manufacturing, far less freight arrived by rail and ship, so waterfront employment also declined. Matthew Drennan estimates that various types of work directly connected with the waterfront in New York—the wholesale trade, water transportation, trucking, warehousing, and railroads—totaled 412,000 in 1969 but fell to 270,000 twenty years later, down 65.5 percent.[43]

Railroad freight handling in the harbor followed the same pattern, and the railroads employed far fewer workers. The railroads serving the Port of New York lost enormous amounts of freight business with the end of manufacturing: not just in Manhattan, but in the entire metropolitan region. With far less freight to be carried across the Hudson River from New Jersey, the railroads no longer needed the huge rail yards crowding the shore in Jersey City and Hoboken. Along the Manhattan waterfront, the railroads also abandoned their freight depots and piers.

≈ The Container Revolution

Technological revolutions often begin quietly. In Boston, Alexander Graham Bell cried out softly, "Watson, come here. I need you." At Kitty Hawk in North Carolina's Outer Banks, the Wright brothers flew the first powered airplane, with only a few people watching. In New York City, buried on page thirty-nine of the *New York Times* on April 27, 1956, a story described a crane lifting "fifty-eight silvery new trailers aboard a tanker . . . the first commercial application of a new concept in shipping," an understatement if there ever was one.[44] Trucking executive Malcom McLean was fed up with the long waits, high costs, and pilferage on the piers in New York Harbor. He decided to load some of his freight destined for Texas into a container that was 8 feet long, 8 feet wide, and 10 feet tall. A crane hoisted the metal boxes onto one of his World War II–era surplus oil tankers, the *Ideal-X*, bound for Houston. Marc Levinson argues that the shipping container "began an era of dramatic change in the business of handling cargo—and reshaping New York City's economy."[45] He views the container as the equivalent of other transportation revolutions: the canal, the railroad, the steamship, and the airplane.

This particular revolution took place not along the Manhattan waterfront, but across the harbor, in Newark Bay, where the Port Authority had leased Newark's city-owned airport and the adjacent piers. In 1950, New York City mayor William O'Dwyer had asked the Port Authority to conduct a study under the premise that the Authority would take over the city's waterfront, modernize the piers, and run the port independent of the Department of Marine and Aviation. The Port Authority proposed to spend over $114 million on this project—a massive investment, equivalent to over $1 billion today. Once again a firestorm erupted, with loud opposition from all waterfront constituencies. Above all, the union and the city's politicians did not want the independent Port Authority to control the waterfront.

Although the city rejected the proposal, the Authority did not abandon its efforts to improve the port area's maritime facilities. In 1947, it won control of Newark's dilapidated piers. Using the enormous surplus generated by its bridges and tunnels, the Authority decided to rebuild Newark's waterfront. In 1952, construction began on 1,500 feet of new piers, as well as three cargo sheds, with a total storage capacity of 270,000 square feet, on 6 acres of land.[46] Nowhere on the Manhattan waterfront or in Brooklyn was there 6 acres of available space. When completed in 1954, at a cost of $6 million, Port Newark offered the most modern and up-to-date pier facilities in the entire port. An additional advantage of the new piers was that they were situated right next to Interstate 95, part of the country's new interstate highway system. Trucks could now load containers and then drive out the gate and onto I-95. In Manhattan and Brooklyn, a truck leaving the docks had to grind through the impossible bottlenecks on city streets, adding hours before the vehicle reached the highways.

In 1955, New Jersey officials announced a new agreement with the Port Authority: to develop 450 acres of marshland south of Newark, in Elizabeth, New Jersey. Over the next few years, the Authority completed the largest maritime infrastructure project ever undertaken in the port's history. Port Elizabeth would become a container facility and, along with Port Newark, the busiest container port on the East Coast.

Although the construction in Port Elizabeth had only recently commenced, the *Ideal-X* loaded its containers and set sail out of Newark

Bay, through the Narrows, and down the Atlantic coast to Houston, Texas. The container revolution had begun. In attendance for the sailing of the *Ideal-X* were the mayor of Newark, the chairman of the Port Authority, and the operators of Port Newark, along with 100 guests. Absent were the mayor of New York City, the commissioner of the Department of Marine and Aviation, the ILA's leaders, and New York shipping executives. In their collective worst dreams, they could not have imagined the whirlwind to follow: the complete destruction of Manhattan's maritime waterfront, which, for over two centuries, had been the leading port in the country and the source of the city's economic vitality.

Back in New York City, the longshoremen, union leaders, shipping executives, Department of Marine and Aviation officials, and political leaders struggled to grapple with the rapid changes along the waterfront. Freight volume slowed as fewer cargo ships arrived at the Manhattan and Brooklyn piers. Transatlantic airline service began to compete with the luxury ocean liner business. Once jet airlines entered the picture, the ships' transatlantic passenger business all but disappeared, leaving the Chelsea piers idle.

While the Port Authority completed the rebuilding of the Newark waterfront, the Department of Marine and Aviation continued to produce elaborate plans to revitalize New York City's waterfront. In 1956, the department asked the city to fund $41 million in improvements.[47] Included were thirteen projects along the Hudson River, with new piers and railroad freight depots from Pier 13, north of the Battery, to Pier 92, at West 52nd Street. The most expensive proposed project involved a new Pier 40, at the foot of West Houston Street in Greenwich Village, to serve both passenger and cargo ships for the Holland-America Line. On the East River, the department's plans envisioned completely rebuilding a 1.75-mile section of the waterfront.

Year after year, the Department of Marine and Aviation, city officials, and New York City's mayor announced that the Manhattan waterfront was about to rebound, despite all evidence to the contrary. In 1963, the Waterfront Commission reported that the hiring of longshoremen along the Hudson waterfront, from the Battery to Chambers Street, had declined by 24.1 percent from 1958 to 1962, and between Chambers and West 12th Streets, by 47 percent.[48] Over in Port Newark and Port Elizabeth, hiring had increased by 50 percent. In the face of the bleak data on waterfront employment in New York City a few years later, Mayor John Lindsay told a meeting of shipping and port executives that the decline had halted: "I think we may have seen the worst. . . . In the past year we have turned the corner."[49] He added that the city planned to construct new container facilities on Staten Island, at Stapleton and at Holland Hook.

Levinson estimates that in 1956, containers carried less than 1 percent of all cargo in the port; twenty years later, they carried 37 percent. The volume of containerized cargo only accelerated over the next decades. As a result, work disappeared on the Manhattan waterfront. The Waterfront Commission's annual reports document the sweeping decline in labor. In 1963–1964, Manhattan's waterfront shipping industry employed 1.4 million days of longshoremen's labor, but by 1975–1976, a mere decade later, that number shrank to 127,041 days.[50] In 1965, 10,313 registered longshoremen worked the port. The Waterfront Commission's FY 2012–2013 report lists the number of licensed dockworkers in the entire port at 5,962, identifying 1,994 as "deep sea" longshoremen, who man the giant cranes lifting the containers without ever touching the cargo inside.[51] An additional 2,104 employees perform warehousing and maintenance work.

A way of life had disappeared from the Manhattan waterfront. Longshoremen and dockworkers living in the immediate neighborhood could not find work. A trip across the river to Port Newark or Port Elizabeth meant an hour of travel each way and no guarantee that they would be hired. Many did not even own a car. Nor could waterfront workers turn to the factories or warehouses in their neighborhoods, since most manufacturing nearby had shut down for good.

Anger among the longshoremen led to threats to strike. In 1956, Nick Simonetti, who worked and lived in the Chelsea neighborhood for twenty-two years, described the area as "ripe for a strike." A coworker, William Lovett, reported that his last check for a week's work totaled $72.64, adding, "You can't live if you have a family."[52] Lovett argued that each longshoreman had lost at least $100 a week in wages because of the decline in the number of ships using the piers in Chelsea. At a boisterous meeting of the ILA at Union Hall, on West 46th Street, the longshoremen declared war on those responsible for the waterfront's demise, proclaiming, "We are fighting for the survival of the Manhattan waterfront . . . and will show that Manhattan's longshoremen are no longer going to take it sitting down."[53] These workers were the early victims of the cataclysmic change to a post-industrial economy.

The ILA recognized the threat containerization posed. At first the union demanded that longshoremen unload and then reload every container arriving in the port. When the shipping companies fought that demand, the ILA, after more strikes, won a settlement that required shippers to pay a fee to the union for all containers moving through the port. In turn, the union would provide a guaranteed yearly income to all registered longshoremen, whether or not they worked. Given the aging of the workforce and the arrangement not to add any new workers unless the demand for labor increased, the shippers agreed to the settlement, realizing that, over time, the number of longshoremen would decrease dramatically. With relative labor peace, the container revolution continued. Nothing New York's mayor, the Department of Marine and Aviation, or the union could do would reverse the demise of the Manhattan waterfront as a place of maritime commerce.

Given the enormous savings in labor costs, the container revolution surged ahead. In 1921, the report of the Port and Harbor Development Commission documented the labor-intensive process of loading and unloading cargo ships. The commission selected a sample of ships using the port and then sent staff to conduct clockings: observing the longshoremen and recording the time they spent working at the various tasks. To unload a transatlantic ship with 2,537 tons of cargo required 4,268 man-hours of labor. Then it took 3,570 man-hours to load 1,752 tons of new cargo. A coastal freighter arriving with 1,114 tons of cargo utilized 1,638 man-hours, and then it was loaded with 996 tons of freight, taking 998 man-hours.[54] Longshoremen still climbed down into the holds of ships and muscled the cargo onto slings, which would lift it out of the hold, or they waited for cargo to be lowered and then stowed it safely for the voyage. For almost 300 years, shipping remained a labor-intensive industry, until shipping containers arrived. The containers drastically reduced the cost of loading and unloading, the most expensive part of shipping. Giant cranes (fig. 8.7), lifting about 30 containers an hour, replaced thousands of longshoreman man-hours.[55] In 1921, unloading 2,537 tons of cargo required 4,268 man-hours of labor. Today that amount of freight can be moved in 98 containers and unloaded by one crane on a giant new container ship in just over three hours. A mod-

Figure 8.7. A container crane in operation. In 1919, a transatlantic cargo ship arriving in New York with 2,537 tons of cargo required 4,268 man-hours of long-shoremen's labor to unload. That amount of cargo today would fill 98 containers and, with the fastest cranes, be unloaded in about 3 hours. *Source*: Port Authority of New York and New Jersey

ern container ship with 6,000 or 8,000 containers is loaded and un-loaded without a single longshoreman down in the hold.[56]

As the container revolution continued, the maritime commerce in the Port of New York abandoned the Manhattan waterfront and moved to Port Newark and Port Elizabeth in New Jersey. Container ports re-quire enough surrounding area to store thousands of containers, space the Manhattan waterfront never had. The Port Authority continues to invest in the container facilities in Newark Bay, with over $4 billion in improvements currently underway. In order to serve the next genera-tion of supersized container ships, the Authority dredged the Kill van Kull, the strait leading from the Upper Bay to Newark Bay. Even with the deeper channel, however, the newer ships cannot fit under the Bay-onne Bridge. The Port Authority's solution is to raise the bridge! The original arch bridge stands 156 feet above high water, but the newest generation of container ships using the port will be over 175 feet in

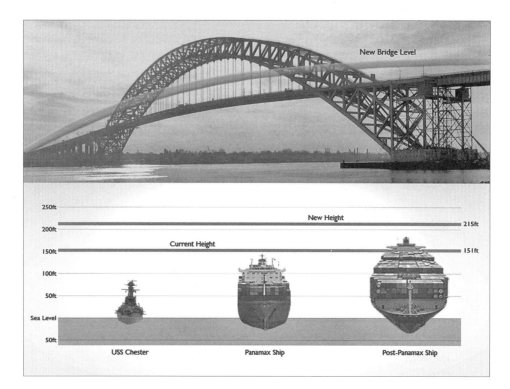

Figure 8.8. Schema for raising the Bayonne Bridge. When raised, the new bridge now allows the largest container ships in the world to enter Newark Bay and berth at the Port Authority's main container terminals in Port Newark and Port Elizabeth, New Jersey. *Source*: Port Authority of New York and New Jersey

height (fig. 8.8). The $1.4 billion project, completed in 2017, raised the new arch 215 feet above high water.[57]

Today, the Port Newark–Elizabeth Marine Terminal is visible from the New Jersey Turnpike. It fills the horizon with a sea of containers, giant gantry cranes, and massive ships that are capable of carrying thousands of containers. In the distant background, the Manhattan skyline towers over the original Port of New York, no longer the center of maritime commerce.

Since the day the *Ideal-X* sailed in August 1956 with fifty-eight small metal boxes, the number of containers has continued to grow dramatically: from 1990 to 2016, the amount handled has increased by more than 225 percent, to over 3.6 million (table 8.2). With the raising of the

Table 8.2.
Port of New York and New Jersey, containers and tonnage

Year	No. of containers	Metric tons (million)
1956 (*Ideal-X*)	58 (10 ft. length)	
1990	1,100,840	
1995	1,311,960	
2000	1,769,000	
2005	2,800,007	28.1
2010	3,076,395	32.2
2014	3,342,286	35.4
2016	3,602,508	36.8

Sources: Port Authority of New York and New Jersey, *Annual Report*, for the years 1990, 1995, 2000, 2010, 2014, and 2016, http://corpinfo.panynj.gov/pages/annual-reports/?_ga=2.62644644.982920952.1502452151-80699361.1498239720 [accessed May 2016].
Note: Percentage increase, 1990–2016 = 227%.

Bayonne Bridge, the Port Authority forecasts continued growth into the future. Experts predict that the next generation of huge container ships coming to Newark Bay could take business away from the country's two largest container ports: Los Angeles and Long Beach, California. Currently, imports from Asia (primarily from China), bound for the East Coast, are unloaded in California and hauled by trains across the United States to the New York metropolitan market, the richest in the country. With the already-widened Panama Canal and the higher Bayonne Bridge, it may become cheaper to send containers bound for the Port of New York directly from China to Newark Bay.

≈ The Death of the Manhattan Waterfront and the Near Death of New York

On the Manhattan waterfront, ships disappeared, as did lighters, car floats, and ferries. Trucks now carry freight over and under the Hudson River, and the need to float freight across the river evaporated. All of the investment by the Department of Marine and Aviation in new pier facilities came to naught. By the 1970s, the proud piers lining the East and Hudson Rivers—where the packet ships once set sail for Liverpool, and the transatlantic steamships for Europe—stood abandoned (fig. 8.9). Some of the piers burned or collapsed into the Hudson. The elevated West Side Highway, over West Street, was literally

Figure 8.9. Abandoned Manhattan piers, circa the 1970s. *Source*: *New York Times* / Redux Pictures

falling apart. It added to the gloom, as did the empty warehouses on the streets leading down to the waterfront.

Decay on the waterfront was only one element in the dramatic decline of New York City during the three decades from the mid-1970s through the 1990s. With the collapse of its industrial base and maritime business, tens of thousands of jobs vanished. Racial change brought tension to the city. A second Great Migration began when millions of African Americans seeking a better life left the South for the "warmth of other suns," and many moved to New York.[58] Seeking economic opportunity and freedom from oppressive racism, they arrived in the city just as its economy entered a three-decade period of change and decline.

At the end of World War II, the lure of the suburbs drew middle-class New York City residents to Nassau, Suffolk, and Westchester Counties, as well as to the suburban communities in northern New Jersey and Connecticut. New York City's population reached 7.8 million in 1960 but then dropped to just under 7 million by 1980. People fled whole neighborhoods. On street after street, abandoned buildings added to the sense of decay. On October 5, 1977, President Jimmy Carter visited

Charlotte Street in the South Bronx and stood in a vacant lot, surrounded by destruction that resembled Berlin at the end of World War II.[59] The city's property tax revenue had plummeted. The rates of street crimes soared, and the crack-cocaine epidemic heightened the fear visitors and ordinary New Yorkers felt walking the streets or riding the graffiti-splashed subway system.

Two years before Carter's visit, the city of New York stood on the edge of bankruptcy. Mayor Abraham Beam's advisors prepared a statement for him to read at a press conference, which was to announce, "I have been advised by the comptroller that the city of New York has insufficient cash on hand to meet debt obligations due today."[60] Beam never issued the statement, however. A coalition of leading citizens, politicians, and union leaders, with the assistance of the New York State Legislature, set up the Municipal Assistance Corporation and the Emergency Financial Control Board, which took control of the city budget. After a vicious political battle in Washington, DC, the federal government extended $1.3 billion in loans to New York City. (The city paid these loans back, with interest, within three years.) As the crisis unfolded, President Gerald Ford threatened to veto any bill providing "a federal bailout of New York City." The next day, the *New York Daily News* published its famous headline: "Ford to City: Drop Dead."[61]

[9]

Rebirth of the Waterfront

Slowly, and with great effort, New York City recovered from the collapse, abandonment, and despair of the 1970s and 1980s, although the legacy of that period proved to be daunting. The crumbling maritime infrastructure along the Hudson and East Rivers would never be rehabilitated. Commercial use of the waterfront would never return. Newark Bay now served as the heart of the Port of New York, where the Port Authority had built one of the most modern container facilities in the world. The empty piers from another century, where generations of longshoremen labored, seemed to have no value whatsoever. No longer would freight from all over the world, cotton from the American South, or coal and iron ore from Pennsylvania cross from water to land. On the adjacent blocks, hundreds of factories and warehouses stood abandoned. Yet the Manhattan waterfront still remained. The death of New York City's maritime world provided an opportunity for rebirth: to reimagine a waterfront that offered views of the Upper Bay and rivers surrounding the island, paths on which to walk and run, and an array of parkland to enjoy.

Once again, the question of public versus private enterprise and control took center stage. Would the rebirth of the Manhattan waterfront be directed and funded by the city of New York, or would the city turn to private ownership and capital to revitalize the shoreline? New York City's history provides a precedent for both courses of action. From 1686 until 1870, the city ceded control of the waterfront to private interests, who constructed and maintained facilities to serve the city's maritime commerce. When private enterprise failed to keep pace with both the flood of ships in the harbor and technological changes in the industry, the Department of Docks repurchased the waterfront and ran the port as a public enterprise.

A third alternative also existed, however: a public authority. Public authorities are institutions created by state governments. They exercise delegated political power and are run by appointed—not elected—executives. Many authorities have the legal right to issue tax-exempt bonds to finance projects and then keep the revenues that the bridges, tunnels, and airports generate. If the facilities generate a "profit" beyond expenses, then the authority has funding that can be used for new projects. An authority and its appointed board decide which developments to proceed with, and then obtain formal approval from elected officials.

Public authorities carry out crucial governmental functions. On the one hand, these entities act as local governments; on the other, they stand apart from the political system that created them, often wielding immense power without direct oversight or control. In New York State, there were 33 authorities in 1956, and 640 in 2004, ranging from the mighty Port Authority of New York and New Jersey to the Tuckahoe Parking Authority in suburban Westchester County.[1] Public authorities built highways, bridges, and new infrastructure in the New York metropolitan region. Robert Moses used public authorities to become a preeminent power broker, shaping the future of New York City and the entire region.[2] Moses never won an election, yet numerous governors of the state and mayors of New York City feared his power and seldom opposed his plans.

By the 1970s, the port and region's first authority, the Port Authority, played an oversized role in the metropolitan region, but by then it was not alone. The Metropolitan Transportation Authority (MTA), the Triborough Bridge and Tunnel Authority, and myriad other large and small authorities have financed, built, maintained, and managed major components of the region's crucial transportation infrastructure. For the rebirth of the waterfront, elected officials and private-interest groups at both the state and local levels turned to public authorities. Just as it had done in the eighteenth and nineteenth centuries, the city of New York again surrendered public control of an incredibly valuable resource: the waterfront surrounding Manhattan Island.

≈ Early Efforts by Public Authorities

Redevelopment of the Manhattan waterfront began with plans to build new commercial space and housing at the southern tip of the island. Piers on the Hudson River, all the way from the Battery to Reade

Street, no longer served either the shipping companies or the railroads and fell into disuse. In the early 1960s, two massive renewal projects began, which changed the waterfront in Lower Manhattan: the World Trade Center and Battery Park City. Neither the city, nor the State of New York, nor private developers built the Trade Center and Battery Park City; public authorities constructed both.

The Port Authority of New York and New Jersey, a public authority created in 1921 to improve the railroad infrastructure serving New York Harbor, announced plans for the World Trade Center in 1961. How a public authority, created to solve the transportation problems in the port, could go into the commercial real estate business in Lower Manhattan is a complicated tale.[3] Plans for the Trade Center's twin office towers included 10 million square feet of office space. At the time, no private real estate developer would undertake such a massive project. Real estate interests complained that with a surplus of vacant commercial space in Lower Manhattan, they would never be able to compete with the tax-subsidized Trade Center. They also predicted that the Port Authority would never be able to lease all of the space.

≈≈ A Massive New Water-Lot

Excavation for the World Trade Center commenced in 1966, to the west of Greenwich Street between Cortlandt and Vesey Streets, on made-land created by water-lot grants, some of which were awarded before the Revolutionary War. Building the foundations for the twin towers required the excavation of 1,200,000 cubic yards of fill from that made-land along the Hudson River, posing a challenge: what was to be done with it? Excavated material could be hauled away, but at great expense. Since the adjoining piers on the river had been abandoned, a solution was readily at hand: create a massive new water-lot where the piers once stood. The 96 acres of new land—1 mile long and extending 900 feet out into the Hudson River—became the site for Battery Park City. The project created the largest single parcel of made-land in the city's history. This new water-lot culminated a process of expanding the island out into the surrounding waterways, one that began in 1686. In 1776, Greenwich Street fronted the shoreline. The waterfront edge of the Battery Park landfill is one-third of a mile farther out into the Hudson River, at Reade Street (map 9.1). Plans for Battery Park City envi-

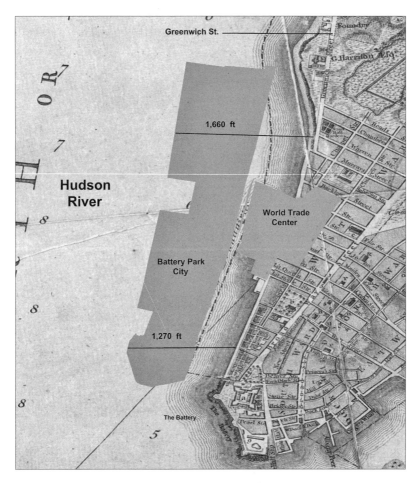

Map 9.1. A 1776 map of the Hudson River and Lower Manhattan, with the footprints of Battery Park City and the World Trade Center superimposed. Battery Park City stands on the largest water-lot in Manhattan's history. *Source*: Created by Kurt Schlichting. Source GIS layers: 1776 Ratzer map, New York Public Library, Map Warper; city lots for Battery Park City and the World Trade Center, New York City Planning Department.

sioned a new community, located where railroad freight and ships from all over the world had once arrived (fig. 9.1).

Battery Park City includes upscale residential and commercial developments, built not by a private real estate company, but by a new public authority, the Battery Park City Authority (BPCA). The origin of this plan began in 1961 with David Rockefeller, who was the president of the

Figure 9.1. The Battery Park City water-lot. Almost all of the piers from the Battery to Jay Street were not in use. Therefore, rather than load the excavated material from the foundations for the World Trade Center onto barges and then tow it to a distant landfill, the Port Authority created an enormous landfill in the Hudson River, adjacent to the twin towers. This new water-lot totaled over 90 acres. *Source*: Battery Park City Authority

Chase Manhattan Bank, the chairman of the Downtown Lower Manhattan Association (DLMA), and the brother of Nelson Rockefeller, the governor of New York. Concerned with the decline of the Wall Street business area and the rise of Midtown Manhattan as the center of the city's commercial activity, the DLMA and the Rockefellers supported the Port Authority's plans for the World Trade Center and the Battery Park City landfill. The DLMA argued that any downtown revitalization efforts must include housing, a departure from focusing on new office space.

Numerous problems in financing the massive project arose, and the first apartments were not completed until 1980. Battery Park City's combination of offices, upscale housing, stores, and venues for leisure activities became a blueprint for the "new urbanism" of the 1980s, encouraging high-density redevelopment in cities as an antidote to suburban sprawl.[4] Aaron Shkuda points to Battery Park City as an exemplar of that trend and of the evolution of New York City's service economy, since "these residences were to serve as a company town for

Wall Street, a new form of housing that would appeal to employees and keep them close to the office around the clock . . . as the New York economy was transforming to a post-industrial service system prior to the exponential growth of Wall Street after deregulation in the 1980s."[5]

This new housing would be for the affluent, however, not the poor. In the 1930s, with the abandonment of the East River shipyards from Corlears Hook to 14th Street, the city of New York built thousands of subsidized apartments for the working class and the poor along the East River. Manhattan's new waterfront renewal, despite pledges to the contrary, now includes only expensive condominiums and apartments and little in the way of affordable housing. The rebirth of the waterfront would serve multiple uses—for commercial space, housing, parkland, and recreation—but these would be mainly for the affluent in a post-industrial economy.

Battery Park City today is home to over 13,000 residents, and 36 acres of the development remain as open space. The market-rate rent for two-bedroom apartments here ranges from $5,600 to over $10,000 a month. The StreetEasy website lists a number of two-bedroom condominiums for sale, at over $2 million each.[6] The most recent census data estimates that the average household income for Battery Park City residents is $316,718, far above Manhattan's average of $132,839.[7] A key part of the plan for this area included an esplanade and marina on the Hudson River, creating a shoreline park for the public (fig. 9.2),

Figure 9.2. The Battery Park esplanade and the Hudson River shoreline. Battery Park City includes a park, which is open to the public, along the newly created waterfront from the Battery to Jay Street. Battery Park serves as the southern anchor for the stunningly successful Hudson River Park. *Source*: Battery Park City Authority

maintained by the BPCA with revenue generated by its real estate developments. This park, designed for public use, was paid for and controlled by a public/private authority, not by the city of New York.

∼∼ Planning and Building a New Waterfront

In 1992, the city of New York published a new waterfront blueprint, officially recognizing the transformation that was taking place. Mayor David Dinkins had directed the New York Department of City Planning to undertake a major study of the waterfront—not for Manhattan alone, but for the city's entire 578-mile shoreline. The long-range plan trumpeted that it "balances the needs of environmentally sensitive areas and the working port with opportunities for waterside public access, open space, housing, and commercial activity."[8] To deal with multiple constituencies spread throughout the metropolitan area's five boroughs, the plan included commercial activity, along with increased public access and the federally required preservation of the natural environment. During the previous three centuries, the public had had little access to the water, and hardly any open space could be found along the lower Manhattan waterfront.

The 1992 plan envisioned a twenty-first century waterfront. It identified a number of priorities, with the first three focused on the natural environment. Maritime activity was relegated to fourth place. New York City's waterfront would become a place where

- parks and open spaces, with a lively mix of activities, are within easy reach of communities throughout the city;
- people once again swim, fish, and boat in clean waters;
- natural habitats are restored and well cared for; and
- maritime and other industries, though reduced in size from their heyday, thrive in locations with adequate infrastructure support.[9]

For the first time in New York City's history, an official plan acknowledged the stunning decline of the maritime world that once drove the city's economy. A priority for the new waterfront was to set goals for the use of the shoreline as parkland, both for recreation and for some restoration of the natural environment. While the shore that Henry Hudson viewed in 1609 could never be restored, every effort would be made to mitigate hundreds of years of maritime exploitation of the waters surrounding Manhattan Island. Conspicuous by their absence, however, were details of how this restoration would be financed.

The plan's focus on restoration followed in the footsteps of a truly remarkable environmental victory: the defeat of Westway, a federally financed highway project along the Hudson River.[10] A high-powered coalition of politicians, labor unions, construction companies, and real estate moguls had used their collective political power to push through plans to replace the crumbling West Side Highway with a new superhighway. It would be built out into the river, in a deep concrete cut that then would be covered over, creating acres of new waterfront to be developed. Initially, the proposal seemed to generate almost universal support from the city's power structure. The federally funded highway, requiring no expenditure by New York City, would bring billions of dollars to a city starved for investment and create thousands of high-paying construction jobs.

An improbable coalition of political activists, the Sierra Club, and local community leaders fought tooth and nail to defeat the plan. In a climactic legal battle, the Sierra Club sued the US Army Corps of Engineers in federal court. The Corps of Engineers had to comply with the legal requirement to conduct an environmental impact study for the proposed highway, which was mandated by the national Environmental Protection Act (1970) and the Clean Water Act (1972). The Corps of Engineers' study initially found that Westway posed no environmental risk, especially to the striped bass population in the Hudson River. The Sierra Club's experts disagreed. In 1985, Thomas P. Griesa, chief judge of the US District Court for the Southern District of New York, ruled in favor of the Sierra Club, and Westway died. The *Village Voice*'s Tom Robbins, celebrating the twenty-fifth anniversary of the defeat of Westway in 2012, summed up the victory:

> We never built Westway, the crazed multi-billion-dollar-city-in-the-river landfill project that Presidents, governors, and mayors desperately fought to build back in the 1980s. You don't remember this? Count your lucky stars. It was one of the last great attempted public arm-twistings by the Permanent Government—a bid to give the ever-campaign-generous real estate industry its most coveted desire: More Manhattan land on which to build. This week marks the twenty-fifth anniversary of one of the great citizen vic-

tories of our time: The defeat of Westway. A band of creative activists, aided by attorneys dedicated to public service, and backstopped by a thoughtful and brave federal judge managed to beat the entire power structure of the city at its own game.[11]

New York did find a silver lining in the project's defeat, however, since changes to the federal highway laws allowed the city to "trade-in" the US government's initial funding of over $1.7 billion for Westway and use it to rebuild the city's crumbling subway system. Without the Westway dollars, the rebirth of the subway would have been delayed for decades.

The Westway fight stimulated alternative plans. One of them imagined the shoreline as public space—a place to remind people daily that Manhattan is an island, surrounded by waterways. The waterfront, no longer gritty and dangerous, would become a place where the public could enjoy magnificent views and watch the sun set over the Hudson River and the Upper Bay. For the wealthy, the waterfront could evolve into a fashionable place to live. For the stylish, it could become a place to shop, dine, and visit an art gallery or museum. Thus, despite the Westway defeat, the shorefront remained contested space. Would people from all walks of life and income be welcome on the revitalized waterfront, or would it become a place apart, for only the rich and successful to enjoy?

Hudson River Park

Battery Park City's promenade created the southern anchor for a new waterfront park to line the Hudson River, extending from the Battery to Riverside Park and then on to Inwood Hill Park, at the northern tip of Manhattan Island. The defeat of Westway and the removal of the elevated West Side Highway presented an opportunity to build a major section of a greenway around the entire island of Manhattan.[12]

In 1992, Governor Mario Cuomo and Mayor David Dinkins signed an agreement to have the city and the state join in funding the new park. Six years later, Governor Pataki signed legislation to create a new public authority, the Hudson River Park Trust, to manage the building of the park. The trust would oversee day-to-day operations and be funded by revenues generated within the park, to be used only for park purposes. This new authority would be self-financed by activities in the park, including a large parking garage on Pier 40 at Houston Street. The trust, not the city's Parks Department, would run the park, a return to the concept of private, rather than public, development. Control rested with an appointed, not elected, governing board.

Over the next two decades, the trust spent $350 million to build the riverfront park. This new parkland rests on the bulkhead constructed by the Department of Docks in the latter half of the nineteenth century. The massive granite stones that cap the bulkhead are visible in sections of the walkway on the river. It is hard to imagine that the park would have been built if a new bulkhead had to be constructed, given the cost involved. As it was, the Hudson River Park Trust, in the face of economic challenges, formed a partnership with the Friends of Hudson River Park, a nonprofit organization founded to promote the building of the park. The latter's mission, as a fundraising partner for the trust, is to seek private philanthropic support for the completion of the park and for its continued maintenance. Neither the state nor the city of New York provides ongoing financial support or directly manages the Hudson River Park. The Parks Department, part of New York City's government, plays a very limited role in the new 550-acre park along the Manhattan waterfront. Instead, a public authority and a private philanthropic organization rule the park, with limited oversight from elected officials at both the local and state levels.

The crucial question of public versus private control remains contentious, however. A case in point occurred when the 2007 fiscal crisis strained the trust's budget, and the organization ran deficits, which reached $9.5 million in 2014.[13] Pier 40, at Houston Street, had become a major problem for the trust. It was built for the Holland-America Line by the Department of Marine and Aviation in the late 1950s, in a last-ditch attempt to save the shipping business on the Manhattan waterfront. The pier survived, and it became part of Hudson River Park, being used for parking and as a sports facility. The Pier 40 parking garage provided the trust with a major source of revenue: $6 million a year. Inspections later found that pilings supporting the pier had deteriorated and needed to be replaced, at an estimated cost of $100 million, well beyond the fiscal resources of the trust and the Friends of Hudson River Park.

One solution was for the city of New York and the state to finance the repairs for Pier 40. The original enabling legislation specifically allows that "additional funding by the state and the city may be allocated as necessary to meet the costs of operating and maintaining the park."[14]

Figure 9.3. Pier 40, on the Hudson River waterfront. The concrete building inland, across West Street, is St. John's Terminal, originally the southern terminus of the High Line. *Source*: Axion Images

Public funding would have come with increased public oversight, however, so instead, the trust turned to private sources of capital. In 2014, the trust quietly negotiated a deal with the Atlas Capital Group, a real estate development company, to transfer development rights from Pier 40—in essence, the air rights—to the other side of West Street, in exchange for $100 million.[15] Across the street, Atlas owns the St. John's Terminal building, built as a rail-freight terminal at the High Line's southern end. In the 1930s, freight trains traveled down the High Line to the terminal to deliver goods to Lower Manhattan. Since the demise of the High Line as a freight line, the five-story terminal building has served as a warehouse. UPS is currently the major tenant.

The Atlas Group manages a real estate portfolio valued at $2.5 billion and proclaims expertise in "repurposing and repositioning existing properties."[16] In exchange for the $100 million, Atlas plans to use the air rights to build high-rise luxury housing where the St. John's Terminal currently stands (fig. 9.3). This new, multistory construction would join numerous other luxury buildings that now line the Hudson River waterfront, extending from Chelsea to Greenwich Village, SoHo, and down to Battery Park City.

Massive reconstruction on the streets adjacent to the waterfront, now filled with luxury housing and renovated brownstones, symbolizes another dramatic change in the waterfront: the Irish neighborhoods have vanished. The few remaining longshoremen work at the vast container facilities in Newark Bay. Waterfront bars have disappeared or found new life as trendy upscale places with historical pictures on the walls, a nod to a working-class past.

High Line Park

One of the most iconic symbols of the rebirth of the waterfront, High Line Park (often referred to simply as the High Line), is not actually on the waterfront. After thirty years of protracted legal battles, in 1929 the New York Central Railroad and the city and state reached an agreement to remove the railroad's tracks from 10th and 11th Avenues, eliminating Death Avenue and the West Side problem. At great expense—which was shared equally by the railroad, the city, and the state—the tracks were relocated to an elevated viaduct through the middle of the city blocks from 34th Street to St. John's Terminal. The High Line included a number of sidings running directly into warehouses and factory buildings on the waterfront (fig. 9.4).

Figure 9.4. The High Line, the New York Central Railroad on the West Side. The tracks passed through the huge Manhattan Refrigeration Company's cold storage building on Gansevoort Street, seen in the background. The elevated railroad tracks have now been transformed into an urban park. *Source*: Digital collections, New York Public Library

The High Line opened in 1934, in the midst of the Great Depression and shortly after the advent of the truck and highway revolution. When the Holland Tunnel opened in 1927, hundreds of trucks used the tunnel each day to haul freight back and forth to New Jersey, eliminating the need to float freight across the Hudson River to the railroad depots on the waterfront. The New York Central's traffic on the High Line also declined. By the 1970s, the elevated tracks were overgrown by weeds and filled with debris, a stark symbol of the decline of the railroads and the City of New York.

Against all odds, and amid calls for the city to tear down the abandoned rail line, two West Side residents, Joshua David and Robert Hammond, formed the Friends of the High Line in 1999 and set out to save the elevated tracks. This was primarily a private initiative.[17] They argued that the abandoned rail line could be redeveloped as an elevated park, modeled after the Promenade plantée in Paris. To build the elevated parkland, David and Hammond raised private funds for the project, eventually amassing over $150 million. The city initially contributed $50 million, and the first section opened in 2009, to rave reviews. Today, the High Line is one of the most visited tourist attractions in New York City.

Mayor Bloomberg officiated at the opening of the second section of the High Line on June 7, 2011. He argued that the total of $115 million contributed to it by the city was a sound investment. The success of the High Line, he continued, had led to over $2 billion in private investment, pointing to "the deluxe apartment buildings whose glass walls press up against the High Line and the hundreds of art galleries, restaurants, and boutiques it overlooks."[18]

The High Line is a few blocks from Hudson River Park, and the two reinforced the transformation of the neighborhoods along the shore. A crowning jewel came with the announcement by the Whitney Museum of American Art that it would relocate to a new building at the southern end of the High Line, at Gansevoort Street. Designed by architect Renzo Piano, the new museum building stands between the High Line and Hudson River Park. Visitors can now walk along the High Line from West 34th Street to Gansevoort Street, visit the Whitney Museum, and then cross to Hudson River Park for a walk to Battery Park City—a journey through a new cityscape that epitomizes the rebirth of the waterfront.

Gentrification of the Irish Waterfront

Not only did High Line Park become a popular tourist destination, but the flood of people visiting it and the numerous art galleries and restaurants in Chelsea have also furthered the gentrification of a working-class waterfront. The city's decision to rezone the blocks inland from the Hudson River opened the former warehouses and factory buildings in Chelsea and Greenwich Village for redevelopment. Warehouses were converted into art galleries, high-end apartments, and luxury condominiums; restaurants and upscale retailers followed.

In Greenwich Village, on the streets running from Hudson Street down to the river, where generations of Irish longshoremen lived, wealthy buyers have purchased the iconic brownstone buildings. In the 1870s, these brownstones were carved up into four or five apartments each, but many have now been converted back to single-family homes. The block on Bethune Street between Washington and Greenwich Streets retains its nineteenth-century appearance, with classic brownstones lining both sides of the street. For example, 32 Bethune Street (see map 9.2), with its exterior beautifully restored, has three floors, with steps down to the basement. The current estimated market value of this Bethune Street residence is over $5 million. Only the truly wealthy can now afford a restored brownstone, located a block and a half from Hudson River Park, in an area of Greenwich Village that once was replete with tenements for impoverished immigrants.

In the late nineteenth century, when much of the shipping in the port moved to the Hudson River, Bethune Street became a working-class neighborhood, where the men toiled on the waterfront. In 1880, two families lived at 32 Bethune Street: Joseph Hayes, an Irish immigrant; and Thomas McDonald, with his wife, three young children, and an Irish maid (table 9.1). By 1900, the McDonalds and Hayes had moved on, and two new families resided there. Mary Bailey and her sister Annie both worked to support their three siblings; and James O'Toole, who came from Ireland in 1870, worked down the street as a longshoreman.

The Irish longshoremen and the working-class residents on Bethune Street are long gone, as is the close-knit surrounding neighborhood that Jane Jacobs celebrated in her classic work, *The Death and Life of Great American Cities*.[19] On Bethune and the other Greenwich Village streets adjacent to the Hudson, the gentrification has been stunning. The eight

Table 9.1.
Census data, 32 Bethune Street

Name	Age	Nativity	Occupation
1880 census			
McDonald, Thomas	35	New York	iceman
McDonald, Mary	36	New York	keeping house
McDonald, Thomas	6	New York	at school
McDonald, Nellie	5	New York	
McDonald, Edward	3	New York	
Dooley, Mary	20	Ireland	servant
Hayes, Joseph	35	Ireland	iceman
1900 census			
Bailey, Mary	28	New York	packer
Bailey, Annie	16	New York	cracker baker
Bailey, Maggie	13	New York	at school
Bailey, James	12	New York	at school
Bailey, Joseph	7	New York	at school
O'Toole, James	48	Ireland	longshoreman
O'Toole, Susan	39	Ireland	
O'Toole, Anna	14	New York	at school
O'Toole, James	11	New York	at school
1910 census			
Kerriochan, Hugh	65	Ireland	carpenter
Kerriochan, Andrew	31	New York	syrup maker
Westley, Agnes	28	New York	sales in dept. store
Westley, Edison	10	New York	
Shay, Annie	45	Ireland	servant
Silver, William	21	New York	lacquerer
Silver, Mrs. W. J.	22	New York	
Silver, Florence	1	New York	
1920 census			
Allen, John	47	Ireland	laborer
Ginnetty, Joseph	47	Ireland	longshoreman
Vonhasselen, Otto	59	Germany	rigger
Vonhasselen, Delia	46	Germany	
Sneezy, Carrie	49	New Jersey	housework
Sneezy, Jesse	13	New Jersey	

Sources: "New York City, Enumeration Districts," United States Federal Census for the years 1880, 1900, 1910, and 1920, www.ancestry.com [accessed spring 2016].

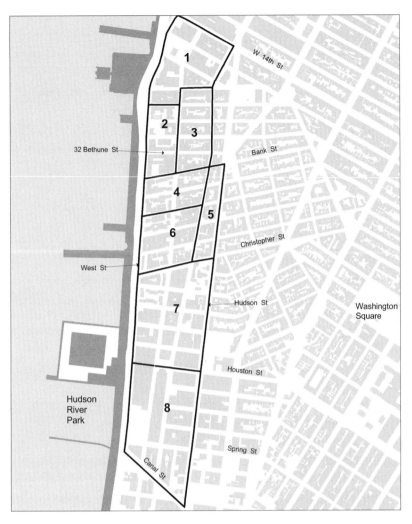

Map 9.2. Hudson River neighborhoods in Greenwich Village. *Source*: Created by Kurt Schlichting. Source GIS layer: building footprints, New York City Planning Department. Other source: census block groups, US Bureau of the Census, "2014 American Community Survey (ACS)," American Factfinder, https://factfinder.census .gov/faces/nav/jsf/pages/searchresults.xhtml?refresh=t/ [accessed March 10, 2016].

neighborhoods (designated as census block groups) between West 14th and Canal Streets (map 9.2) are home to a population significantly wealthier, whiter, and better educated than the average city resident (table 9.2). Not surprisingly, these neighborhoods have an average household income well above the average for Manhattan.

Table 9.2.
Greenwich Village population, by block groups, in 2014.
The block group numbers refer to the areas shown on map 9.2.

Location	Population	Average income ($)	White ethnicity (%)	College+ education (%)
Manhattan		132,938	60.0	59.3
Block group 1	1,171	283,283	90.1	71.8
Block group 2	1,235	124,825	94.4	84.5
Block group 3	1,147	184,387	91.5	85.9
Block group 4	1,638	232,629	89.9	82.2
Block group 5	1,175	123,666	75.7	74.4
Block group 6	1,741	149,752	86.7	81.8
Block group 7	2,684	222,065	76.7	86.2
Block group 8	821	178,148	67.8	70.2

Source: US Bureau of the Census, "2014 American Community Survey (ACS)," American Fact-finder, https://factfinder.census.gov/faces/nav/jsf/pages/searchresults.xhtml?refresh=t/ [accessed March 10, 2016].

The Meatpacking District, the southern terminus of the High Line (map 9.2, #1), has household incomes that are more than two times the Manhattan average. The district is filled with luxury condominiums, fancy restaurants, upscale stores, and trendy bars, just as Mayor Bloomberg had envisioned. In 1992, the city's plan for a revitalized waterfront proposed using the shoreline as a place for all to gather, but now only the rich can afford to live nearby.

The rebirth of the waterfront has not been limited to Greenwich Village. On Manhattan's West Side, from 38th Street in Chelsea south to the Battery, the number of people living along the shore has grown dramatically from the depths reached in the 1970s. The 1970 census reported that 8,063 people lived in this 3.7-mile section along the Hudson River. By 2014, the population had risen to 28,597, an increase of over 250 percent.[20] As new apartments and condominiums are completed, and with hundreds more on the drawing boards, the number of people on the West Side waterfront will only increase. No longer do the men go down to the docks for the shape-up, as the Irish did. Rather, both the men and women living there earn high salaries in the city's post-maritime, post-industrial economy.

≈≈ Public versus Private Control Redux

The pressure by private interests to build on and develop the waterfront remains unabated. The creation of the Hudson River Park Trust as a private/public partnership proved to be a bold move to revitalize the riverfront, and the requirement that the park be self-financing defied the accepted logic of a public park. Yet with no continuing public financial support, the park had no alternative but to turn to private interests to save Pier 40. The trust, in return for the needed funds, had to trade a valuable asset: expansive views of the Hudson River.

The proposed development around Pier 40 has drawn much criticism. The Atlas Capital Group's plans for five residential buildings include commitments to set aside a portion of the 1,500 planned new apartments as affordable housing. At a public hearing in November 2015, New York State Assembly member Deborah Glick, who represents the area, raised the issue of how "affordable" would be defined. A *New York Times* editorial called the project a "win-win" for the West Side of Manhattan, but it did not directly address the issue of who would have enough money to rent an apartment or buy a condominium there.[21] In 2016, the average condominium in Manhattan sold for $2 million. If the ten penthouses in the proposed buildings sell for $10 million each, the trade-off is certainly a win for their private developer, who will easily cover the $100 million in pier renovations and can then profit from the remaining 1,490 apartments. Time and again in recent years, developers promise to include affordable housing, but citizens of modest means remain priced out or have to use the "poor door" in the back of the building.[22]

Another controversial private proposal further illustrates the contemporary battle over control of the waterfront. Following private negotiations with the Hudson River Trust in 2014, Barry Diller and his wife, Diane von Furstenberg, whose philanthropy supported the High Line, announced plans to fund the construction of a 2.4-acre island north of Pier 55, at West 13th Street.[23] The proposed island park would serve as a venue for concerts and the arts. Diller and von Furstenberg pledged $113 million to privately fund the project, which immediately set off another storm of controversy. Environmental groups objected. Critics charged that the public would have to pay park entrance fees to attend performances on the new island, which would be connected to Hudson River Park, although the trust gave assurances that parts of the island would be free for all to visit, even when concerts were scheduled.

In a *New York Times* editorial entitled "The Billionaires' Park," David Callahan argues that this controversy illustrates the city's abandonment of its responsibility to fund public spaces for all to enjoy freely. In the nineteenth century, New York City led the world by providing public parks and public institutions that served as models of civic responsibility: Central Park, Bryant Park, and the New York Public Library. Callahan claims: "Private wealth is remaking the city's public spaces. . . . While it's hard to argue with more parks, or the generosity of donors like Mr. Diller, this isn't just about new patches of green. It's more evidence of how a hollowed out public sector is losing its critical role, and how private wealth is taking the wheel and having a growing say over basic parts of American life."[24] The city of New York has abnegated its responsibility and left the rebirth of the waterfront in the hands of public authorities and private philanthropy.

Opposition to the proposed island continued, led by a small but loud group of community activists and secretly funded by Douglas Durst, a New York real estate billionaire. After two years of legal battles, with the cost estimate rising to over $250 million, Mr. Diller announced on September 13, 2017, that "Because of huge escalating costs . . . it was no longer viable for us [referring to his wife] to proceed."[25]

The apparent end of the Diller island plan echoes the battle over Westway in the 1970s. New York's power structure, including Governor Andrew Cuomo, Senator Chuck Schumer, Mayor Bill de Blasio, and the local community board all supported the plan. A *New York Times* editorial criticized the opponents of the plan as part of a now "familiar dynamic in New York when the well-heeled put their money behind a new idea: an assumption that something untoward is afoot, including privatization," yet added, "Private money is not necessarily to be shunned."[26] The *Times* editorial does not offer guidance on how to balance the public good against private interest, an issue that has embroiled the use of the Manhattan waterfront for hundreds of years.

≈≈ The Waterfront in an Era of Climate Change

Private development continues unabated on the made-land along the Hudson and East Rivers, with one grandiose plan following another.

Despite its high real estate prices, Manhattan's waterfront has been transformed and made accessible. Anyone who knew New York in the 1970s could not have imagined the greenway that now almost entirely encircles Manhattan Island. Once a wasteland of abandoned piers, crumbling highways, and empty warehouses, the shoreline is today a place where millions enjoy the waterways that surround the island. A dangerous, dirty, and polluted shoreline has been reclaimed for recreation and open space.

One inexorable consequence of the relentless development and redevelopment of the Manhattan waterfront over time remains: the waterfront that Henry Hudson and the early Dutch settlers saw has literally disappeared. As map 9.3 illustrates, the made-land along the Manhattan shoreline today extends out from the original shore into the Hudson and East Rivers, but it is only a few feet above sea level. Thus this redevelopment of the waterfront has ignored potential environmental challenges. As Hurricane Sandy reminded the city, the waters that surround Manhattan—once and still a source of wealth—may now, in an era of global warming, prove to be a major threat.

The flooding from Superstorm Sandy inundated Lower Manhattan, putting it under as much as 4 or 5 feet of water, resulting in billions of dollars in damages. Along both rivers, the water covered almost all of the made-land that was created by filling in water-lots over the centuries, including the largest water-lot of all: Battery Park City. In many areas of Lower Manhattan, the shoreline drawn on the 1776 Ratzer map marks the high-water line of Sandy's flooding and coincides with the Federal Emergency Management Agency's (FEMA) current flood zones.

Hurricane Sandy engendered $18 billion in damages. In New York City, forty-three people died. Water covered the open space in Battery Park City and flooded the buildings along the East River from Water Street to the river's edge, including the ferry buildings at the Battery and the South Street Seaport. The Brooklyn-Battery Tunnel flooded, as did the subway tunnels connecting Manhattan to Brooklyn. The L line at 14th Street—between Manhattan and Williamsburg, in Brooklyn—will be closed for eighteen months, beginning in January 2019, to repair the flood damage from Sandy. Disaster also struck the Con Edison plant at 14th Street, which exploded, plunging Lower Manhattan into darkness for days without electricity. The Con Ed generating plant sits

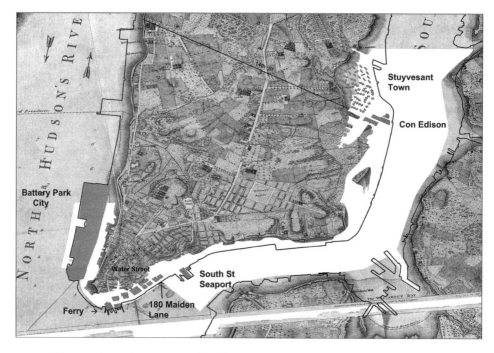

Map 9.3. A 1776 Ratzer map of Manhattan Island, with a black line showing the modern shoreline. The shaded area represents the current FEMA flood zone. In an era of climate change, the made-land on the Manhattan waterfront remains vulnerable. Hurricane Sandy flooded the Con Edison plant on 14th Street, plunging the entire southern part of the island into darkness. On the Wall Street waterfront, hundreds of millions of dollars of commercial property sits on what was once the East River. *Source*: Created by Kurt Schlichting. Source GIS layers: 1776 Ratzer map, New York Public Library; shoreline, lots, and building footprint maps, New York City Planning Department; floodplain map, Federal Emergency Management Agency.

on a water-lot created in the 1830s. Moreover, half of Stuyvesant Town and Peter Cooper Village stands on the made-land, which was once Stuyvesant Cove, clearly visible on the 1776 Ratzer map.

The relentless development along the rivers on water-lot land has left billions of dollars of New York City real estate vulnerable to climate change and storms of Sandy's intensity. For example, 180 Maiden Lane, built on made-land on the East River, is one of fourteen office buildings between the Battery and Maiden Lane. It is a forty-one–story Class A building, built in 1982, that contains over 1 million square feet of office space. Hurricane Sandy flooded the basement, forcing tenants to find alternative office space for months, until the water damage was repaired. In January 2015, the building, valued at $435 a square foot, sold for $470,000,000—almost half a billion dollars for one office building on the corner of Maiden Lane and South Street.[27] The new owners of 180 Maiden Lane must reckon with the risk of future storms.

The fourteen nearby buildings between the Battery and Maiden Lane total 13,657,595 square feet, and their commercial value is over $5.9 billion (at $435 a square foot). All of these buildings sit on made-land from some of the first water-lots on the East River, and all were flooded by Sandy. They stand in the FEMA flood zone for Lower Manhattan. The risk, moreover, is not limited to this one part of the Manhattan shore. It includes all of the made-land created over hundreds of years.

As New York City plans for the future, the threat of rising sea levels and violent storms is a problem that will have to be dealt with pragmatically. The question of control of the waterfront persists, and it complicates efforts to address global warming. The city, primarily through zoning and building codes, exercises only limited regulatory power over development on the waterfront. The federal government's FEMA regulations do deal with the issue of building in a flood zone, but existing property is grandfathered in.

On the Staten Island and Long Island shorelines devastated by Hurricane Sandy, both the federal government and the state's government have recommended buying out private owners of the most vulnerable shoreline property. If a similar approach were to be used for Manhattan property lying in the FEMA floodplain, staggering questions would arise. Where would the billions of dollars required to purchase the commercial and residential property on made-land around Manhattan come from? Where would the private businesses and luxury residential housing relocate to? Another option is to build enormous dikes and walls, as the Netherlands has done. Yet dikes around Manhattan Island would destroy access to the city's waterfront.

≈≈ The Future of the Waterfront

Mayor Michael Bloomberg and the Department of City Planning published *Vision 2020* in March 2011, five months before Hurricane Irene and a year and a half before Hurricane Sandy. Mayor Bloomberg's foreword presents a clear mandate: to continue to "reclaim one of our most valuable assets—our waterfront."[28] The report hailed the work done to date and listed priorities for going forward: "These projects are part of the most sweeping transformation of an urban waterfront in American history. And as we serve as an engine of economic growth for America and the world, we will invest in initiatives that will help New Yorkers green their communities and build a more economically sustainable city. Jobs and the environment, affordable housing and open space, waterborne transportation and in-water recreation—all these priorities have informed *Vision 2020*'s goals for our City's more than 550 miles of shoreline."[29]

The *Vision 2020* plan did not anticipate the ferocity of Hurricanes Irene and Sandy. For instance, at the Battery on the night of October 29, 2012, at the height of Sandy, the water rose 14 feet above low water—much higher than the bulkhead around Lower Manhattan—flooding the residential and business districts, with their private property on made-land. To protect the city in the future requires setting priorities to save the waterfront from rising storm waters and establishing the respective responsibilities of both the city and private-property owners.

A year after Superstorm Sandy, Mayor Bloomberg unveiled a $20 billion plan to protect New York from the next hurricanes.[30] The 450-page report, *A Stronger, More Resilient New York*, includes detailed proposals to strengthen the shoreline of the entire city.[31] The study identified the consequences of the historic expansion of the shoreline, "with areas along these coasts now lying at or near the water level. Examples of these low-lying areas include Lower Manhattan."[32] All of the valuable properties along the rivers are vulnerable, including 180 Maiden Lane, the proposed St. John's Terminal development, and the controversial Diller–von Furstenberg island.

Among the plans considered in this report are giant storm barriers across the Narrows and the East River, to be closed when major storms approach. The model for these barriers is the massive storm gates the Dutch have built to protect their country, which lies below the level of the North Sea. Such barriers, however, would take years to be approved, cost between $20 and $25 billion, and require millions of dollars for annual maintenance. The Bloomberg plan did not recommend building the barriers, arguing instead for an "integrated flood protection system" for Manhattan Island and other vulnerable shorefronts.

The integrated plan includes "deployable floodwalls," with moveable posts and panels to be installed when storms threaten, and then to be removed after the storms pass. No cost estimates are provided, but the report identifies operating challenges: storage, heavy equipment to erect the posts and panels, a sizable workforce, and regular emergency practice drills.[33] Whatever plan is eventually implemented, it will be incredibly controversial and expensive.

A glittering cityscape has arisen on the once gritty-waterfront of Manhattan Island. The Atlantic World of commerce, based on shipping and the railroads, that generated the original prosperity of New York City has been eclipsed by the world of finance and wealth. New fortunes have completely transformed the neighborhoods where millions of immigrants first settled. On the new waterfront, most of the redevelopment has been carried out by private capital, not by the city of New York. As with the use of water-lots to create a maritime infrastructure in the eighteenth and nineteenth centuries, the city has once again ceded the future of the waterfront to a powerful combination of public authorities and private interests. It remains to be seen whether this complex arrangement of private/public control will meet the environmental challenges to come.

Notes

1 Growth, Decline, and Rebirth

1 Ted Steinberg, *Gotham Unbound: The Ecological History of Greater New York* (New York: Simon & Schuster, 2014), appendix F, 365.

2 Robert Albion, *Square-Riggers on Schedule: The New York Sailing Packets to England, France, and the Cotton Ports* (Princeton, NJ: Princeton University Press, 1938).

3 William Cronin, *Nature's Metropolis: Chicago and the Great West* (New York: W. W. Norton, 1991).

4 Joseph Goldstein, "Along New York Harbor 'On the Waterfront' Endures," *New York Times*, January 8, 2017.

2 Water-Lots and the Extension of the Manhattan Shoreline

1 See Jill Lepore, *Encounters in the New World: A History in Documents* (New York: Oxford University Press, 2002); Douglas Egerton, Alison Games, Jane G. Landers, Kris Lane, and Donald R. Wright, *The Atlantic World: A History, 1400–1888* (New York: Wiley Blackwell, 2007).

2 Matthew Fontaine Maury, *Explorations and Sailing Directions to Accompany the Wind and Current Charts* (Washington, DC: C. Alexander, 1851), 254–255.

3 Albion, *Square-Riggers*, appendix 2, 276–277.

4 Robert Albion, *The Rise of New York Port, 1815–1860* (New York: Charles Scribner's Sons, 1939), 16.

5 There are numerous books that chronicle the building of the Erie Canal and its importance in both American history and the history of the rise of New York City. See, for example, Peter Bernstein, *Wedding of the Waters: The Erie Canal and the Making of a Great Nation* (New York: W. W. Norton, 2005); Ralph Andrist, *The Erie Canal* (Rockville, MD: American Heritage, 2016); Gerard Koeppel, *Bond of Union: Building the Erie Canal and the American Empire* (Cambridge, MA: Da Capo Press, 2009); Evan Cornog, *The Birth of Empire: DeWitt Clinton and the American Experience, 1769–1828* (New York: Oxford University Press, 2009).

6   Albion, *Rise of New York Port*, 29.

7   Richard MacKay, *South Street: A Maritime History of New York* (New York: G. P. Putnam & Sons, 1934).

8   US Bureau of the Census, "Population in the Colonial and Continental Period," chap. 1 in *A Century of Population Growth, from the First Census of the United States to the Twelfth, 1790 to 1900* (Washington, DC: US Government Printing Office, 1909), 11.

9   Albion, *Rise of New York Port*, 117.

10 Hendrik Hartog, *Public Property and Private Power: The Corporation of the City of New York in American Law, 1730–1870* (Ithaca, NY: Cornell University Press, 1983).

11 *Charter of the City of New York*, Internet archive, https://archive.org/details /charterofcityofnoonewy/ [accessed August 14, 2015].

12 Ibid.

13 Hartog, *Public Property and Private Power*, 14.

14 Corporation of the City of New York, *Minutes of the Common Council of the City of New York*, vol. 4, *January 24, 1730 to September 19, 1740* (New York: Dodd Mead, 1905), 101–105.

15 Ibid., November 18, 1731, 104.

16 Ibid., 105.

17 Ibid., 102.

18 Ibid.

19 George William Edwards, *New York as an Eighteenth-Century Municipality*, vol. 2, *1731–1776* (New York: Columbia University Press, 1917), 51–52, Internet archive, https://archive.org/details/newyorkaseighteeooedwaiala/ [accessed May 2015].

20 Hartog, *Public Property and Private Power*, 34.

21 Edwards, *New York*, 158.

22 Ibid.

23 M#31, Carlyle Street, 1799, New York Municipal Archives [hereafter NYMA].

24 Sidney Hoag Jr., "The Dock Department and the New York Docks," paper no. 16, presented March 22, 1905, *Municipal Engineers of the City of New York, Proceedings for 1905*, 38.

25 M#30, water grants, 1730, Department of Ports and Trade [Department of Docks], folder 30, NYMA.

26 Steinberg, *Gotham Unbound*, 57.

27 Ann Buttenwieser, *Manhattan Water-Bound: Manhattan's Waterfront from the Seventeenth Century to the Present* (Syracuse, NY: Syracuse University Press, 1999), 43.

28 "Welcome to New York, 1609," Welikia: Beyond Mannahatta, https://welikia.org /explore/mannahatta-map/ [accessed March 2015].

29 M#82, 15th to 23rd Streets, 10–12, 1825, NYMA.

30 Hartog references criticism of the water-lot grants that had appeared in an early newspaper, the *Independent Reflector*, in 1753. See Milton Klein, ed., *The Independent Reflector; or, Weekly Essays on Sundry Important Subjects; More Particularly Adapted to the Province of New-York* (Cambridge, MA: Harvard University Press, 1959), 118–127.

31 See New York Commissioners of the Sinking Fund, *The Wharves, Piers, and Slips, Belonging to the Corporation of the City of New York, 1868*, vol. 1 (New York: New York Printing, 1868), 9.

3 The Ascendency of the Port of New York

1 State of New York, *Report of Commissioners Relative to Encroachments in the Harbor of New York*, Senate, No. 3 (Albany: C. Van Benthuysen, 1857), 5.

2 Buttenwieser, *Manhattan Water-Bound*, 129.

3 John H. Morrison, *History of the New York Shipyards* (New York: W. F. Sametz, 1909), 92–93.

4 See David Scott, *Leviathan: The Rise of Britain as a World Power* (London: Harper Press, 2013).

5 Albion, *Square-Riggers*, 6. Albion's exhaustive research with the New York and Liverpool shipping records provides data not available from any other source.

6 Ibid., title page.

7 Ibid., 15.

8 *Gazette and Commercial Advertiser*, November 15, 1816, n.p.

9 Cathy D. Matson, *Merchants and Empire: Colonial New York* (Baltimore: Johns Hopkins University Press, 1998), appendix A, 320–321.

10 David Davis, "James Crooper and the British Anti-Slavery Movement, 1821–1823," *Journal of Negro History*, vol. 45, no. 4 (October 1960), 244–247.

11 Albion, *Square-Riggers*, appendix 14, 322.

12 "Liverpool Customs Bill of Entry No. 2009," July 16, 1825, Merseyside Maritime Museum, Archives, and Library, Liverpool, England [hereafter Merseyside Museum].

13 Eugene R. Dattel, *Cotton and Race in the Making of America: The Human Costs of Economic Power* (Lanham, MD: Bowman & Littlefield, 2009), x.

14 Ibid., 29.

15 Albion, *Square-Riggers*, 20.

16 Between 1818 and 1856, just over a hundred different packet ships provided service between New York and Liverpool: fifty to London and fifty-one to Le Havre, France. See Albion, *Square-Riggers*, appendix 2, 275–286.

17 State of New York, *Encroachments*, 14.

18 US Department of Commerce, *Statistical Record of the Progress of the United States, 1800–1914, and Monetary, Commercial, and Financial Statistics of Principal Countries* (Washington, DC: US Government Printing Office, 1914).

19 Albion, *Square-Riggers*, 83–87.

20 Ibid., appendix 17, 326.

21 Morrison, *New York Shipyards*, 54.

22 Albion, *Square-Riggers*, appendix 19, 330.

23 US Bureau of the Census, "New York City, Ward 1," in *Population Schedules of the 7th Census of the United States, 1850* (Washington, DC: US Government Printing Office, 1852), 130–168.

24 "1850 United States Federal Census, New York Ward 1, Eastern Division," 168, www.ancestry.com [accessed February 2015]. The 1850 census does not include building addresses, but its records contain data for each of the "dwelling houses numbered in the order of visitation" and a second number for "families numbered in order of visitation." A family included all individuals living together. A residential building or tenement starts with a dwelling number, and then the separate families/households in the building or tenement each have a separate "family" number.

25 New York City Landmarks Preservation Commission, *American Seamen's Friend Society Sailors' Home and Institute*, LP-2080, November 28, 2000, 2.

26 "Sailors on Shore," *Sailor's Magazine*, vol. 10, no. 3 (November 1837), 78.

27 "Fatal Fracas in a Bar-Room," *New York Times*, August 13, 1861, 5.

28 "1860 United States Federal Census, New York, Ward 4, District 1," www.ances try.com [accessed July 29, 2016]. The 1860 census does not include addresses, just the ward and district in New York.

29 "Found Drowned," *New York Times*, August 13, 1861, 5.

30 "Murder of a Sailor," *New York Times*, December 11, 1860, 5.

31 "Are Sailors More Depraved Than Formerly?," *Sailor's Magazine*, vol. 31, no. 1 (September 1858), 11.

32 "Shameful Case of Kidnapping," *New York Times*, December 1, 1857, 5.

33 Ibid.

34 "No Advance Wages," *New York Times*, July 24, 1857, 5.

35 Joseph Conrad, *The Nigger of the "Narcissus": A Tale of the Sea* (New York: Doubleday, 1914), 191.

36 Albion, *Square-Riggers*, 49.

37 Ibid., appendix 2, 286–294.

38 Ibid., 51.

39 Sven Beckert, "The Rise of a Global Commodity," chap. 1 in *Empire of Cotton: A Global History* (New York: Knopf, 2014), 3–28.

40 David Cohn, *The Life and Times of King Cotton* (Westport, CT: Greenwood Press, 1973), 88, quoted in Dattel, *Cotton and Race*, 30.

41 Walter Johnson, *River of Dark Dreams: Slavery and Empire in the Cotton Kingdom* (Cambridge, MA: Harvard University Press, 2013), 31.

42 US Bureau of the Census, "Statistics of Slaves," in *A Century of Population Growth from the First Census of the United States to the Twelfth, 1790–1900* (Washington, DC: US Government Printing Office, 1909), 131.

43 Johnson, *River of Dark Dreams*, 41.

44 James Watkins, *King Cotton: A Historical & Statistical Review, 1790 to 1908* (New York: J. L. Watkins & Sons, 1908).

45 The English industrial revolution destroyed the cottage-industry production of thread and the weaving of cloth in northwestern England. The weavers and their families had no choice but to leave their rural cottages, move to Birmingham and Manchester, and work in the new mechanized textile mills. The entire family had to work from dawn to dusk to survive; they had no choice.

46 US Department of Commerce, Bureau of Foreign and Domestic Commerce, *Statistical Record, 1800–1914*, table 370.

47 Dattel, *Cotton and Race*, 86, 85.

48 John Killick, "Risk, Specialization, and Profit in the Mercantile Sector of the Nineteenth-Century Cotton Trade: Alexander Brown and Sons, 1820–80," *Business History*, vol. 16, no. 1 (1974), 2.

49 "Ledger A, 1825–1829," Brown Brothers & Company Papers, Manuscript Division, New York Public Library [hereafter NYPL].

50 Beckert, *Empire of Cotton*, 204.

51 Ibid.

52 "New Orleans Office," vol. 124, 10, Brown Brothers & Co. Papers, NYPL.

53 Ibid., 40.

54 "Gold Account 33," General Ledger, 118–121, Brown Brothers & Co. Papers, NYPL.

55 E. J. Donnell, *Chronological and Statistical History of Cotton* (New York: James Sutton, 1872).

56 Albion, *Square-Riggers*, appendix 8, 307.

57 "Liverpool Customs Bills of Entry," September 1828 and September 1855, Merseyside Museum. This is also the source for the information in the table below.

	No. of ships	No. of bales	Percentage of bales	No. of ships	No. of bales	Percentage of bales
From		1828			1855	
New York	14	9,828	25.2	8	11,431	22.5
Charleston	16	12,305	—	—	—	—
Mobile	—	—	—	6	17,725	—
Savannah	7	5,843	—	—	—	—
New Orleans	11	10,034	25.7	8	19,610	38.6
other	4	985	—	2	3,056	—
Totals	52	38,995		24	50,826	

58 Henry Smithers, *Liverpool, Its Commerce, Statistics, and Institutions: With a History of the Cotton Trade* (Liverpool, UK: Thos. Kay, 1825), 174.

59 Ibid., 171–172.

4 New York's Waterway Empires

1 See Glenn Knoblock, *The American Clipper Ship, 1845–1920* (Jefferson, NC: McFarland, 2014); Richard McKay, *Donald McKay and His Famous Sailing Ships* (New York: Dover, 1995).

2 See Tom Lewis, *Hudson: A History* (New Haven, CT: Yale University Press, 2007).

3 *New York Evening Post*, October 1, 1807.

4 Gibbons v. Ogden, 22 U.S. 1 (1824).

5 John H. Morrison, *History of American Steam Navigation* (New York: W. F. Sametz, 1903).

6 See Frances Dunwell, "The Foundry at Cold Spring," chap. 5 in *The Hudson River Highlands* (New York: Columbia University Press, 1981).

7 J. H. French, ed., *Gazetteer of the State of New York* (Syracuse, NY: R. P. Smith, 1860), 57.

8 Ibid., appendix F.

9 Edwin Burrows and Mike Wallace, *Gotham: A History of New York City from 1898 to 1919* (New York: Oxford University Press, 1999), 654.

10 Morrison, *History of American Steam Navigation*.

11 See Jon Sterngass, *First Resorts: Pursuing Pleasure at Saratoga Springs, Newport, and Coney Island* (Baltimore: Johns Hopkins University Press, 2001).

12 Ibid.

13 US Bureau of the Census, *Manufactures of the United States in 1860: Compiled from the Original Returns of the Eighth Census* (Washington, DC: Government Printing Office, 1865), xxi.

14 Roger McAdam, *Old Fall River Line: Being a Chronicle of the World-Renowned Steamship Line, with Tales of Romantic Events and Personages during Its Ninety Years of Daily Service and Accounts of the Last Voyages by the Famous Sound Steamers* (Brattleboro, VT: Stephen Daye Press, 1937).

15 See Brian Cudahy, *Over and Back: The History of Ferryboats in New York Harbor* (New York: Fordham University Press, 1990).

16 Kenneth Jackson, ed., *The Encyclopedia of New York City* (New Haven, CT: Yale University Press, 1995), 398–400.

17 Ibid., 397–401.

18 Board of Commissioners of the Metropolitan Police, *Annual Report, 1866* (New York: Bergen & Tripp, 1867), 61–64.

19 Ibid.

20 See David McCullough, *The Great Bridge: The Epic Story of the Building of the Brooklyn Bridge* (New York: Simon & Schuster, 1972).

21 Jackson, *Encyclopedia of New York City*, 149.

22 See David Hammack, "Urbanization Policy: The Creation of Greater New York," chap. 7 in *Power and Society: Greater New York at the Turn of the Century* (New York: Russell Sage Foundation, 1982). Another reason for consolidation involved efforts by the Republicans to weaken the power of Tammany Hall, Manhattan's Democratic political machine. See Terry Golway, *Machine Made: Tammany Hall and the Creation of Modern American Politics* (New York: W. W. Norton, 2014).

23 French, *Gazetteer of the State*, 213–223.

24 US Department of Homeland Security, *Yearbook of Immigration Statistics: 2012* (Washington, DC: US Department of Homeland Security, Office of Immigration Statistics, 2013).

25 See John Kelly, *The Graves Are Walking: The Great Famine and the Saga of the Irish People* (New York: Henry Holt, 2012). The literature of Irish immigration to America, both popular and academic, is voluminous.

26 "Famine Irish Passenger Record Data File," in the series "Records for Passengers Who Arrived at the Port of New York during the Irish Famine, Created 1977–1989, Documenting the Period 1/12/1846 to 12/31/1851," National Archives and Records Administration [hereafter Famine Irish database, NARA], https://aad.archives.gov/aad/fielded-search.jsp?dt=180&tf=F&cat=all&bc=sl/ [accessed November 2014].

27 See Tyler Anbinder, *Five Points: The Nineteenth-Century New York City Neighborhood That Invented Tap Dance, Stole Elections, and Became the World's Most Notorious Slum* (New York: Basic Books, 2001).

28 Albion, *Square-Riggers*, 276.

29 Herman Melville, *Redburn: His First Voyage; Being the Sailor-Boy Confessions and Reminiscences of the Son-of-a-Gentleman, in the Merchant Service* (New York: Penguin, 1976).

30 The seventy-six ships arrived on twenty-six different days in July.

31 Famine Irish database, NARA.

32 Ibid.

33 British Customs, Liverpool, "Bills of Entry," September 1855, Merseyside Museum; "New York Passenger List," Oct.–Nov. 1855, www.ancestry.com [accessed

February 2015]. These are also the sources for the information in the table below.

Packets	To Liverpool		To New York	No. of Passengers	
	Date of arrival	No. of bales of cotton	Date of arrival	Cabin	Steerage
Black Ball Line					
Isaac Webb	Sept. 13	1,629	Oct. 30	13	411
Albion	Sept. 17	722	Oct. 31	138	372
Pacific	Sept. 17	822	Oct.	22	276
Red Star Line					
Constellation	Sept. 18	1,304	Nov. 12	91	501
Dramatic Line					
Roscius	Sept. 5	828	Nov. 9	10	91
freighters					
Driver	Sept. 20	1,734	Nov. 14	89	493
E. Z.	Sept. 20	681	Nov. 17	2	19
Totals		7,720		365	2,163

34 For an analysis of the costs for passengers in the nineteenth century, see Raymond Cohn, "The Transition from Sail to Steam in Immigration to the United States," *Journal of Economic History*, vol. 65, no. 2 (June 2005), 469–494.

35 Burrows and Wallace, *Gotham*, viii–ix.

36 US Bureau of the Census, *Report on the Manufactures of the United States at the Tenth Census (June 1, 1880)* (Washington, DC: US Government Printing Office, 1883), xxvi; US Bureau of the Census, *Census Reports, Twelfth Census of the United States, Taken in the Year 1900: Manufactures* (Washington, DC: US Government Printing Office, 1902).

37 US Bureau of the Census, *Report on the Manufactures . . . 1880*, xxvi.

38 Burrows and Wallace, *Gotham*, 659.

39 Jackson, *Encyclopedia of New York City*, 855.

5 The Social Construction of the Waterfront

1 Charles Barnes, *The Longshoremen* (New York: Russell Sage Foundation, 1914), 169–170.

2 Ibid., 168–169.

3 Ibid., viii.

4 Jackson, *Encyclopedia of New York City*, 184–185.

5 Barnes, *Longshoremen*, 5.

6 See Stanley Nadel, *Little Germany: Ethnicity, Religion, and Class in New York City, 1845–1880* (Urbana: University of Illinois Press, 1990).

7 For the source of all 1880 census data, see "NAPP 1880 US Census, 100% Enumeration," Steven Ruggles, Katie Genadek, Ronald Goeken, Josiah Grover, and

Matthew Sobek, *Integrated Public Use Microdata Series: Version 6.0* [database] (Minneapolis: University of Minnesota, 2015), North Atlantic Population Project, https://www.nappdata.org/napp/.

8 See Kurt Schlichting, "'Kleindeutschland,' the Lower East Side in New York City at Tompkins Square in the 1880s: Exploring Immigration Patterns at the Street and Building Level," in Ian Gregory, Don DeBats, and Dan Lafreniere, *Routledge Handbook of Spatial History* (London: Routledge, 2017).

9 Ibid. This is also the source for the information in the table below.

	New York City (Manhattan) (%)	Ward 4 (%)	Ward 7 (%)	Ward 17 (%)	Greenwich Village (%)	Chelsea (%)
Native	18.2	6.7	9.0	8.0	30.5	21.6
"Irish"	34.2	58.6	54.7	17.9	41.5	51.1
Born in Ireland	16.5	28.4	25.1	8.8	18.5	23.5
1st gen. Irish	17.7	30.2	29.6	9.1	23.0	27.6
"German"	28.3	14.9	16.3	58.1	11.2	12.1
Born in Germany	13.3	8.0	7.6	28.5	5.1	5.4
1st gen. German	15.0	6.9	8.7	29.6	6.1	6.7

10 Burroughs and Wallace, *Gotham*, 745.

11 US Bureau of the Census, "New York, Table II—Population by Subdivisions of Counties," *Seventh Census of the United States, 1850* (Washington, DC: Government Printing Office, 1853), 102.

12 Norman Hapgood and Henry Moskowitz, *Up from the City Streets: Alfred E. Smith; A Biographical Study in Contemporary Politics* (New York: Harcourt, Brace, 1927), 5.

13 Ibid., 3–5.

14 Ibid., 5.

15 Citizens' Association, Council of Hygiene and Public Health, *Sanitary Condition of the City* (New York: D. Appleton, 1865).

16 Citizens' Association, "Report of the Fourth Sanitary District," in *Sanitary Condition of the City*, 45.

17 Ibid.

18 Ibid., 47.

19 Richard Plunz, *A History of Housing in New York City* (New York: Columbia University Press, 1990), 8.

20 Citizens' Association, "Report of the Fourth Sanitary District," in *Sanitary Condition of the City*, 52.

21 Ibid.

22 Ibid., 58.

23 Ibid.

24 See Plunz, "Government Intervention," chap. 7 in *History of Housing*.

25 "Tenement Life in New York," *Harper's Weekly Magazine*, March 29, 1879, 246.

26 "NAPP 1880 US Census, 100% Enumeration."

27 Jacob Riis, *The Battle with the Slum* (New York: Macmillan, 1902).

28 Citizens' Association, "Report of the Fourth Sanitary District," in *Sanitary Condition of the City*, 47.

29 Jacob Riis, "The Down Town Back-Alleys," chap. 4 in *How the Other Half Lives* (New York: C. Scribner's Sons, 1890).

30 Riis, *Battle with the Slum*, 19.

31 Ibid., 17.

32 James Fisher, *On the Irish Waterfront: The Crusader, the Movie, and the Soul of the Port of New York* (Ithaca, NY: Cornell University Press, 2009), 3.

33 "NAPP 1880 US Census, 100% Enumeration." Also see the table in endnote 8.

34 Stuart Waldman, *Maritime Mile: The Story of the Greenwich Village Waterfront* (New York: Makaya Press, 2002), 42.

35 Citizens' Association, "Report of the Fourth Sanitary District," in *Sanitary Condition of the City*, section B, 71.

36 Ibid., 119.

37 "NAPP 1880 US Census, 100% Enumeration"; for the 1900–1930 US census data, the NAPP took either a 5 percent or 10 percent sample of all records from each census. These are also the sources for the information in the table below.

	Percentages				
	1880	1900	1910	1920	1930
Native	30.5	18.7	18.7	21.9	29.2
"Irish"	41.5	35.1	35.5	35.4	8.1
Born in Ireland	18.5	20.6	20.6	8.8	3.5
1st gen. Irish	23.0	15.8	15.8	27.0	5.6

38 New York Public Library, Map Warper, http://maps.nypl.org/warper/.

39 Citizens' Association, "Report of the Twentieth Sanitary District," in *Sanitary Condition of the City*, 238.

40 Ibid., 239.

41 Citizens' Association, "Report of the Seventeenth Sanitary District," in *Sanitary Condition of the City*, 197.

42 Ibid., 198.

43 See "London Terrace," Wikipedia, https://en.wikipedia.org/wiki/London_Terrace/ [accessed November 13, 2015].

44 Fisher, *On the Irish Waterfront.*

6 The Port Prospers, the Railroads Arrive, and Congestion Ensues

1 Albion, *Rise of New York Port*, appendix 7, 398.

2 Albion cautions that the data for coastal shipping—drawn from customs-data shipping publications, such as the *Shipping and Commercial List*, and local newspapers—is incomplete. Many small coastal ships did not register with US Customs or were not reported in the newspapers.

3 "Marine Intelligence," *New York Times*, June 23, 1852, 8.

4 The ships came from Virginia, with wood; Philadelphia, with coal, as well as fish and fish oil for the Fulton Fish Market; Salem, Massachusetts, with general merchandise; Machias, Maine, with lumber; and Portland, Connecticut, with tons of stone from the Brainerd Quarry Company, used to construct New York City's signature brownstone buildings.

5 See T. J. Stiles, *The First Tycoon: The Epic Life of Cornelius Vanderbilt* (New York: Alfred Knopf, 2009).

6 Laws of the State of New York . . . 1831, chapters 263, 323.

7 "Detail of the Real Estate Owned in the City of New York as of Dec. 31, 1854," New York & Hudson River Railroad Papers, box 74, Manuscript Division, NYPL. The New York & Harlem Railroad acquired land at 42nd Street in the 1840s and 1850s for a rail yard to service their steam locomotives. At that time, 42nd Street was out in the country, 2 miles to the north of New York City, whose northern boundary was around Canal Street. Eventually the New York & Harlem owned all of the land from 42nd to 58th Streets, between Lexington and Madison Avenues.

8 In the 1840s, the northern boundary of the city was Canal Street. Above it were large farms owned by descendants of the early Dutch settlers, such as the Stuyvesant family. There were also many farms farther up Manhattan Island. The city of New York owned a great deal of property, including much of the land where Central Park would be constructed. First, though, the city had to evict squatters on the Central Park land, including African Americans and Irish living in what was known as Seneca Village.

9 Albion, *Rise of New York Port*, 314.

10 See Charles Henning, *International Finance* (Ann Arbor: University of Michigan Press, 1958), for a general discussion on the history of documents used in foreign trade.

11 Morrison, *History of American Steam Navigation*.

12 Ibid., 413.

13 Jacob Abbott, "The Novelty Works, with Some Description of the Machinery and the Processes Employed in the Construction of Marine Steam-Engines of the Largest Class," *Harper's New Monthly Magazine*, vol. 2, no. 12 (May 1851), 721–734.

14 "The Loss of the *Arctic*," editorial, *New York Times*, October 12, 1854, 4.

15 See Thomas Kettel, "Banking," chap. 4 in *Southern Wealth and Northern Profits* (New York: George W. & John A. Wood, 1860).

16 US House of Representatives, *Causes of the Reduction of American Tonnage and the Decline of Navigation Interests*, 41st Cong., 2d Sess., 1869–1870 (Washington, DC: US Government Printing Office, 1870), appendix 23, 272–273.

17 Knots are the maritime measure of speed through the water: 1 knot equals 1.15 miles per hour. The *Britannia*'s average speed of 8.5 knots equaled 9.8 miles per hour.

18 See John Maxtone-Graham, *The Only Way to Cross: The Golden Era of the Great Atlantic Express Liners—from the* Mauretania *to the* France *and the* Queen Elizabeth 2 (New York: Barnes & Noble, 1997); Philip Dawson, *The Liner: Retrospective and Renaissance* (New York: W. W. Norton, 2006).

19 "Destroying the Commerce and the Harbor of New York," editorial, *New York Times*, April 17, 1857, 4.

20 Ibid.

21 William O. Stoddard, "New York Harbor Police," *Harper's New Monthly Magazine*, vol. 45, no. 269 (October 1872), 672–683.

22 See Goldstein, "Along New York Harbor."

23 "Destroying the Commerce," 4.

24 State of New York, in Senate, January 29, 1857, *Report of Commissioners Relative to Encroachments in the Harbor of New York*, No. 40 (Albany: C. Van Benthuysen, 1857), 3.

25 Ibid., 30.

26 Ibid., 22.

27 Ibid., 261.

28 Ibid., 187.

29 Ibid., 190.

30 State of New York, in Senate, February 5, 1862, *Report of the Select Committee Appointed by the Last Senate to Examine into the Affairs and Investigate the Charges of Malfeasance in Office, of the Harbor Masters of New York*, No. 100.

31 State of New York, in Senate, January 8, 1856, *Report of Commissioners Relative to Encroachments in the Harbor of New York*, No. 3 (Albany: C. Van Benthuysen, 1856), 6.

32 State of New York, *Encroachments*, No. 40, 21.

33 Ibid., 27.

34 New York City Commissioners of the Sinking Fund, *The Wharves, Piers and Slips, Belonging to the Corporation of the City of New York, 1868*, vol. 1, *East River* (New York: New York Printing, 1868), 1.

35 Ibid., 15.

36 "City Budget for 1867," *New York Times*, February 28, 1867, 2.

7 The Public and Control of the Waterfront

1 Laws of the State of New York . . . 1870, chapter 137.

2 Hoag, "Dock Department," 44–47.

3 Department of Docks, *First Annual Report of the Department of Docks for the Year Ending April 30, 1871* (New York: Douglas Taylor, 1872).

4 Hoag, "Dock Department," 40.

5 Department of Docks, *Thirty-Fifth Annual Report of the Department of Docks and Ferries . . . 1905*, 208.

6 Department of Docks, "Record of Wharf Property Acquired," *Thirty-Third Annual Report of the Department of Docks and Ferries . . . 1903*, 165.

7 Department of Docks, *Thirty-Fifth Annual Report*, 86, 205.

8 New York City Department of Finance, "Individual Property Assessment Roll Data," using the "Your Property Information" web page, http://nycprop.nyc.gov /nycproperty/nynav/jsp/selectbbl.jsp [accessed April–May 2015].

9 Department of Docks, "Report of the Engineer-in-Chief," *Fifth Annual Report of the Department of Docks . . . 1875*, 45–47.

10 Hoag, "Dock Department," 142.

11 Department of Docks, *Thirty-Fifth Annual Report*, 98–99.

12 Hoag, "Dock Department," 141.

13 Fisher, *On the Irish Waterfront*, 7.

14 Department of Docks, "List of Those Employed by the Board, April 30, 1887," *Seventeenth Annual Report of the Department of Docks . . .1887*, 9.

15 "Old Dock Board Politics: Other Things Being Equal," *New York Times*, October 29, 1895, 7.

16 Mazet Committee, *Investigation of Offices and Departments of the City of New York by a Special Committee of the [New York State] Assembly, Report of Counsel, December 22, 1899*.

17 "Mazet Committee's Work: Treasurer Murphy Tells of Docks Department," *New York Times*, August 3, 1899, 3.

18 Ibid.

19 New York Chamber of Commerce, *Thirty-Ninth Annual Report*, May 7, 1896, 4.

20 Ibid., 6–7.

21 Ibid., 4–7.

22 See Hoag, "Dock Department."

23 Department of Docks, "Report of the Engineer-in-Chief," *Tenth Annual Report of the Department of Docks . . .1880*, 114–117.

24 Department of Docks, "Record of Wharf Property Acquired by Department since Its Organization to December 31, 1905," *Thirty-Fifth Annual Report*, 199–200.

25 Department of Docks, "Rent-Roll of Leases and Permits," *Thirty-Fifth Annual Report*, 110–111.

26 "Cry for More Dock Room," *New York Times*, December 14, 1908, 12.

27 "A New and Great Work for New York City," *New York Times*, Sunday magazine, December 27, 1908, SM3.

28 "New Chelsea Piers to Open Tomorrow," *New York Times*, February 20, 1908, C6.

29 Department of Docks, *Forty-Eighth Annual Report of the Department of Docks . . . 1919*.

Line	Pier nos.	Leases ($)	
		1910–1920	1930–1940
Cunard Line	53, 54, 56	177,500	239,400
French Line	57	70,000	84,700
Atlantic Transit	58	70,000	84,700
White Star Line	59, 60	140,000	169,400
Red Star Line	61	70,000	84,700
American Line	62	37,500	45,375
Total		565,000	708,275

30 New York New Jersey Port and Harbor Development Commission, *Joint Report, with Comprehensive Plans and Recommendations* (Albany: J. B. Lyon, 1920), 177.

31 Albion, *Rise of New York Port*, appendix 7, 398.

32 Port and Harbor Development Commission, *Joint Report*, 140.

33 William Wilgus, "A Suggestion for an Interstate Metropolitan District," 15–16, box 48, William J. Wilgus Papers, Manuscript Division, NYPL.

34 See William J. Wilgus, "Milestones in the Life of a Civil Engineer," unpublished manuscript, 1948 Linda Hall Library of Science, Engineering & Technology, University of Missouri–Kansas City.

35 William J. Wilgus, *Transporting the A.E.F. in Western Europe, 1917–1919* (New York: Columbia University Press, 1931), 211.

36 Committee on the Regional Plan of New York and Its Environs, *Regional Plan of New York and Its Environs*, 2 vols. (Philadelphia: Wm. F. Fell, 1929).

37 Port and Harbor Development Commission, *Joint Report*, 4.

38 Jameson Doig, *Empire on the Hudson: Entrepreneurial Vision and Political Power at the Port of New York Authority* (New York: Columbia University Press, 2002), appendix, 403.

39 Ibid.

40 John Griffin, *The Port of New York* (New York: City College Press, 1959), 76.

8 Crime, Corruption, and the Death of the Manhattan Waterfront

1 US Bureau of the Census, *Quarterly Summary of Foreign Commerce of the United States, 1960*, no. 1198 (Washington, DC: US Government Printing Office, 1961), 895.

2 Port and Harbor Development Commission, *Joint Report*, 128–130.

3 Ira Place, chief counsel of the New York Central Railroad, in *Report of the Commission to Investigate the Surface Railroad Situation in the City of New York, on the West Side* (Albany: J. B. Lyon, 1918), 202–204.

4 Ibid., 26.

5 Port and Harbor Development Commission, *Joint Report*, 151.

6 Department of Docks and Ferries, "Rent Role of Leases and Permits," *Thirty-Eighth Annual Report of the Department of Docks and Ferries . . . 1909*, 80–81.

7 William J. Wilgus, "Plan for a Proposed New Railroad System for the Transportation and Distribution of Freight by Improved Methods," box 45, folder 12, William J. Wilgus Papers, Manuscript Division, NYPL.

8 Francis Lane, "A Proposed Subway Belt Line for Lower New York," *Engineering News* [New York], vol. 60, no. 16 (October 5, 1908), 419; "Gigantic Plan to Reduce Street Congestion," *New York Times*, Sunday magazine, October 4, 1908, SM10.

9 "Belt Line Plan Recommended for the Port of New York," *Engineering News-Record*, vol. 86 (February 10, 1921), 271–272.

10 Carl Condit, *The Port of New York*, vol. 2, *A History of the Rail and Terminal System from the Grand Central Electrification to the Present* (Chicago: University of Chicago Press, 1981), 221.

11 Stoddard, "New York Harbor Police."

12 Fisher, *On the Irish Waterfront*, 43.

13 Commission on Industrial Relations, *Industrial Relations: Final Report and Testimony Submitted to Congress by the Commission on Industrial Relations Created by the Act of August 23, 1912*, 11 vols., US Senate, 64th Cong. (Washington, DC: US Government Printing Office, 1916), 3:2054.

14 Ibid., 2055.

15 Ibid., 2059.

16 Malcolm Johnson, *On the Waterfront: The Pulitzer Prize–Winning Articles That Inspired the Classic Movie and Transformed the Harbor of New York* (New York: Chamberlain Bros., 2005), 152.

17 Nathan Ward, *Dark Harbor: The War for the New York Waterfront* (New York: Farrar, Straus & Giroux, 2010), 60.

18 Johnson, *On the Waterfront*, 118.

19 Ward, *Dark Harbor*, 116.

20 Johnson, *On the Waterfront*, 146.

21 A. H. Weiler, "Review of 'On the Waterfront': Brando Stars in Film Directed by Kazan," *New York Times*, July 29, 1954.

22 See Robert Coles, *Dorothy Day: A Radical Devotion* (Reading, MA: Addison-Wesley, 1987); Dorothy Day, *The Long Loneliness: The Autobiography of Dorothy Day* (San Francisco: Harper & Row, 1981).

23 See Pope Leo XIII, "Rerum Novorum" [On the Rights of Capital and Labor], May 15, 1891; Pope Pius XI, "Quadragisimo Anno" [On Reconstruction of the Social Order], May 15, 1931, both in Papal Encyclicals Online, www.papalencyclicals.net [accessed January 2016].

24 Fisher, *On the Irish Waterfront*, 108.

25 In 1952, Kazan's anti-Communist testimony as a witness in Senator Joseph McCarthy's House Un-American Activities Committee helped end some Hollywood careers.

26 "Priest Advocates for Union Revolt," *New York Times*, January 12, 1953, K51.

27 James Fisher, "John M. Corridan, S.J., and the Battle for the Soul of the Waterfront, 1948–1954," *US Catholic Historian*, vol. 16, no. 4 (Fall 1998), 71–87; 80.

28 Ibid., 81.

29 Fisher, *On the Irish Waterfront*, 288.

30 Johnson, *On the Waterfront*, 307.

31 Daniel Bell, *The End of Ideology: On the Exhaustion of Political Ideas in the Fifties* (Glencoe, IL: Free Press, 1960), 170–209.

32 Ibid., 207–208.

33 "Filmsite Movie Review, 'On the Waterfront' (1954)," AMC Filmsite, www.filmsite.org/onth.html [accessed November 18, 2015].

34 "Joseph Ryan Quiet Curses," *New York Daily News*, June 3, 1999.

35 Fisher, *On the Irish Waterfront*, 236.

36 Waterfront Commission of New York Harbor, *Annual Report, FY 2012–2013*, 2, www.wcnyh.gov [accessed January 2016]. Also see Goldstein, "Along New York Harbor."

37 Benjamin Chinitz, "A Century of Slowing Down," chap. 2 in *Freight and the Metropolis: The Impact of America's Transport Revolutions on the New York Region* (Cambridge, MA: Harvard University Press, 1960), 18–48.

38 Ibid., table 2, 21.

39 Peter Kihss, "Pier Growth Here Far behind Rivals," *New York Times*, August 13, 1953, 28.

40 Ibid.

41 Department of Marine and Aviation, City of New York, *A Progress Report to Mayor Robert F. Wagner on Rebuilding New York City's Waterfront* (New York: Department of Marine and Aviation, September 5, 1956).

42 Chinitz, *Freight and the Metropolis*, 132.

43 Matthew Drennan, "The Decline and Rise of the New York Economy," in John Mollenkopf and Manuel Castells, eds., *Dual City: Restructuring New York* (New York: Russell Sage Foundation, 1991), 32.

44 "Tankers to Carry 2-Way Pay Loads," *New York Times*, April 27, 1956, 39.

45 Marc Levinson, "Container Shipping and the Decline of New York, 1955–1972," *Business History Review*, vol. 80 (Spring 2006), 49.

46 Charles Zerner, "Big Port Terminal Near Completion," *New York Times*, January 31, 1954, S8.

47 Department of Marine and Aviation, *Rebuilding New York's Waterfront*, 11.

48 "New York's Docks Lose as Jersey's Gain in 4 Years," *New York Times*, May 12, 1963, S14.

49 Tania Long, "Decline of Port Has Halted Lindsay Tells Propeller Club," *New York Times*, February 24, 1967, 70.

50 Marc Levinson, *The Box: How the Shipping Container Made the World Smaller and the World Economy Bigger* (Princeton, NJ: Princeton University Press, 2006), 72.

51 Waterfront Commission of New York Harbor, *Annual Report, FY 2012–2013*, 9.

52 Werner Bamberger, "Chelsea Dock Workers Irked by Decline in Jobs," *New York Times*, June 4, 1967, 88.

53 Werner Bamberger, "Dockers Demand Waterfront Aid," *New York Times*, June 28, 1967, 89.

54 Port and Harbor Development Commission, *Joint Report*, 186.

55 Ibid., 188.

56 Author's interview with Dan Pastore, Port Authority of New York and New Jersey, May 2016.

57 Port Authority of New York and New Jersey, "Bayonne Bridge Navigational Clearance Project," www.panynj.gov/bridges-tunnels/bayonne-navigational-clearance-project.html.

58 See Isabel Wilkinson, *The Warmth of Other Suns: The Epic Story of America's Great Migration* (New York: Vintage, 2010).

59 Lee Dembart, "Carter Takes a 'Sobering' Trip to South Bronx: Carter Finds Hope Amid Blight," *New York Times*, October 6, 1977.

60 Ralph Blumenthal, "Recalling New York at the Brink of Bankruptcy," *New York Times*, December 5, 2002.

61 Frank Van Riper, "Ford to City: Drop Dead," *New York Daily News*, October 30, 1975, 1.

9 Rebirth of the Waterfront

1 Alan G. Hevesi, *Public Authority Reform: Reining in New York's Secret Government* (Albany: New York State, Office of State Comptroller, 2004), 6, 54.

2 See Robert Caro, *The Power Broker: Robert Moses and the Fall of New York* (New York: Knopf, 1974).

3 See Doig, *Empire on the Hudson*.

4 See the New Urbanism website, www.newurbanism.org.

5 Aaron Shkuda, "Housing the 'Front Office to the World': Urban Planning for the Service Economy in Battery Park City in New York," *Journal of City Planning*, vol. 13, no. 3 (2013), 235.

6 "Battery Park City Apartments for Rent," StreetEasy, http://streeteasy.com/for-rent/battery-park-city/beds:2?sort_by=price_desc/ [accessed March 2, 2016].

7 US Bureau of the Census, "2014 American Community Survey (ACS)," American FactFinder, https://factfinder.census.gov/faces/nav/jsf/pages/searchresults.xhtml?refresh=t/ [accessed March 10, 2016].

8 Department of City Planning, City of New York, *New York City Comprehensive Waterfront Plan: Reclaiming the City's Edge* (New York: Department of City Planning, 1992), i.

9 Ibid.

10 See William Buzbee, *Fighting Westway: Environmental Law, Citizen Activism, and the Regulatory War That Transformed New York City* (Ithaca, NY: Cornell University Press, 2014); Buttenwieser, *Manhattan Water-Bound*, 207–214.

11 Tom Robbins, "Westway, the Highway That Tried to Eat New York, Defeated 25 Years Ago This Week," *Village Voice*, October 1, 2010.

12 "Manhattan Waterfront Greenway Bike Map," NYC Bike Maps, www.nycbike maps.com/maps/manhattan-waterfront-greenway-bike-map/.

13 Teresa Agovino, "Hudson River Park Is Still Sinking Financially," *Crain's New York Business*, June 27, 2013.

14 Laws of the State of New York . . . 1998, chapter 592 (S. 7845).

15 Charles Bagli, "Possible Deal May Bring Money to Repair Pier 40 in Manhattan," *New York Times*, May 15, 2014.

16 "About Us," Atlas Capital Group, www.atlas-cap.com/about-us.php [accessed March 6, 2016].

17 See Annik LaFarge and Rick Darke, *On the High Line: Exploring America's Most Original Urban Park* (New York: Thames & Hudson, 2014).

18 Patrick McGeehan, "The High Line Isn't Just a Sight to See: It's Also an Economic Dynamo," *New York Times*, June 6, 2011, A19.

19 Jane Jacobs, *The Death and Life of Great American Cities* (New York: Random House, 1961).

20 Author's analysis of US Bureau of the Census data for the 1970 decennial census and its "2014 American Community Survey."

21 "A Potential Win-Win for the West Side," editorial, *New York Times*, December 14, 2015.

22 See Laura Kusisto, "A 'Poor Door' on a Planned New York Apartment Tower with Affordable Housing Gets a Makeover," *Wall Street Journal*, August 28, 2014.

23 Charles V. Bagli and Robin Pogrebin, "With Bold Park Plan, Mogul Hopes to Leave Mark on New York's West Side," *New York Times*, November 17, 2014, A1.

24 David Callahan, "The Billionaires' Park," editorial, *New York Times*, November 24, 2014, A27.

25 Charles Bagli, "Billionaire Diller's Plan for Elaborate Pier in the Hudson is Dead," *New York Times*, Sept. 13, 2017, A1.

26 Editorial: "How New Yorkers Sank a Floating Park," *New York Times*, Sept. 15, 2017, A18.

27 Lois Weiss, "100 Wall St.'s $275M Pricetag a Downtown Record," *New York Post*, May 19, 2015.

28 Department of City Planning, City of New York, *Vision 2020, New York City Comprehensive Waterfront Plan* (New York: Department of City Planning, March 2011), 2.

29 Ibid.

30 Kia Gregory and Marc Santora, "Bloomberg Outlines $20 Billion Storm Protection Plan," *New York Times*, June 11, 2013.

31 Department of City Planning, City of New York, *A Stronger, More Resilient New York* (New York: Department of City Planning, June 2013).

32 Ibid., 41.

33 Ibid., 54.

Index

Note: Page numbers in *italics* indicate figures and maps.

Ahearn, Thomas, 140
airline service, 2, 13, 155, 156, 180
Albany (NY): Dutch trading post in, 19; Hudson River lines to, 63, 64–65; railroads serving, 130; steamboat service to, 77
Albion, Robert: on cotton trade, 10; on interest groups, 24; on New York, 18, 19; on sailors, 48; on shipping activity, 40, 45, 52, 53, 109, 215n5, 221n2
Allen, John, 202
American Expeditionary Force (AEF), 153
American Federation of Labor (AFL), 173, 175
Appalachian Mountains, 10, 68
Army Corps of Engineers, US, 143–44, 147, 152, 195
Astor, John Jacob, 135, *136*
Atlantic Basin, *67*, 69
Atlantic World, 16–19, 211
Atlas Capital Group, 198, 205

Bachley, Thomas, 77
Bailey, Mary and family, 201, 202
Baltimore (MD), 1, 18, 19; packet lines to, 45, 76; port fees in, 126; shipping activity at, 21, 46, 176–77
Baltimore & Ohio Railroad, *113*, 114, 163
Barco, Augustin, 50
barges: Erie Canal, *66*, 67, 69, 109; Hudson River, 68, 158

Barry, Bridget and family, 81
Battery, The, 19, 208–9, 210
Battery Park, 7, *193*, 196
Battery Park City, 7, 14, 208; pedestrian access to, 200; water-lot for, 190–94, 207
Bayonne Bridge, 183–85
Beam, Abraham, 187
Beard, Edward, 69
Beckert, Sven, 57
Bell, Alexander Graham, 178
Bell, Daniel, 174–75
Bell, Isaac, 45
Bentaus, Henry, 50
Black Ball Line, 40–43; building, 47; cargo of, 44, 59, 219n33; founders of, 8, 41–42, 79; passengers on, 77, 79–82, 110, 219n33; rivals of, 44–45; schedules for, 116; ships of, 41–44, 77, 79, 80–82, 110, 115, 120, *121*, 219n33; speed of, 17–18, 115, 118
Blau, Waldron, 27
Bloomberg, Michael, 200, 204, 209–11
boarding houses, 49, 50–51, 167
Boorman, Henry, 51
Boston (MA), 1, 18; British occupation of, 19; imports at, 46; packet lines of, 45; port fees in, 126; rail service to, 71, 130; shipping activity at, 109, 110, 177; steamboat service to, 71
Boston & Albany Railroad, 112, 130
Boston and Providence Railroad, 71
Bradford, Hezekiah, 32
Brando, Marlon, 12, 170, 173, 175
Bremen (Germany), packet lines from, 79, 119

bridges, 156; Bayonne, 183–85; Brook-
lyn, 75, 93, *94*
Bronx (NY), 75–76, 142–43, 218n22
Brooklyn (NY), 1, 75–76, 142–43,
218n22; ferries to, 29, 30, 74–75;
manufacturing in, 84, 177, 178; pop-
ulation of, 75; waterfront expansion
in, 151
Brooklyn-Battery Tunnel, 14, 207
Brooklyn Bridge, 75, 93, *94*
Brown, Alexander, 56
Brown, James, 56
Brown, William, 56
Brown & Bell shipyard, 47, 117
Brown Brothers, 56–58
Brown Shipley, 56, 57, 58
Brunel, Isambard Kingdom, 116
Bryant Park, 206
bulkheads: construction of, 4–5, 25,
34; leases for, 140, 161, 166; master
plan for, 133–34; parkland resting
on, 197; public reclamation of, 131–
32, 134–37; rebuilding, 129–30, 137–
40, 143, 144, 147–48; regulation of,
125–26. *See also* piers
Bulkley family, 52–53
Burke, Danny, 99
Burke, Peter, 140
Burke, Thomas, 169
Burr, D. H., 31
Burroughs, Edwin, 83

California: container ports in, 185;
gold rush in, 115
Callahan, David, 206
canals, 67, 68–70. *See also* Erie Canal
Carey, Fr. Philip, 171
car floats, 158, *159*, 161
Caribbean, shipping routes to, 8, 76,
110
Carter, Jimmy, 186–87
Casey, James, 100
Catholic Church, 12
Catholic Workers Movement, 171
census data, US: of 1790, 1, 54; Chelsea, 204, 228n20; Gotham Court,
100; Greenwich Village, *105*, 201–4,
221n37; Irish and German residents
in, 91–92, 220n9; metropolitan regional, 154; slave population in, 54–
55; Ward 1, 48–49, 215n24; Ward 4,
50, 93, 95, 216n28

Central Park, 206, 222n8
Central Railroad of New Jersey, *113*,
114, 158–59, 163
Chalmers, Alexander, 51
Champlain Canal, 67, 70, 78
Charleston (SC), 18, 21; cotton from,
9, 47, 217n57; packet lines to, 52–53,
54, 76; shipping activity at, 109, 110
Chelsea, 7; ethnic makeup of, 91, *92*,
102, 106–8, 220n9; gentrification of,
201; population of, 204, 228n20;
water-lots in, 35–37; workforce in,
168, 181
Chelsea piers, 36–37, 106; decline of,
176; leases for, 150, 224n29; rebuild-
ing, *134*, 146–51, *148*, *149*
Chicago (IL), waterway to, 10, 62, 68,
77
China, trade with, 39, 109, 185
Chinitz, Benjamin, 176
climate change, 14–15, 206–9
Clinton, DeWitt, 18, 68
coal, 32–33, 70, 84
Collins, Edward, 117, 118, 119
Collins, Israel, 117
Collins, Philip, 50
Common Council, 4; Department of
Docks and, 132; laws published by,
23; port control issue and, 128, 131;
water-lot grants by, 27, 29, 34
Conrad, Joseph, 52
Consolidated Edison, 14–15, 33, 207–8
Consolidated Gas Company of New
York, 33, 106
Constitution, US: Fifth Amendment
of, 137; interstate commerce clause
of, 66
container revolution, 13, 180–81, 199
container shipping, 2, 178–85; bridge
height for, *184*; New Jersey, 12, 13,
156, 177, 183–84, 188, 199
Cooke, Bishop Terrence James, 174
Cooper, Peter, 95
Corridan, Fr. John (Pete), 12, 171, 172,
173–74, 175
Costello, John, 101
cotton, 54–58; clothing made from,
85–86; export value of, 46–47, 55;
New York–Liverpool packet lines
for, 9–10, 40, 43, 44, 55, 59, 61, 82,
110, 217n57, 219n33; production of,
44, 54–55, 56, 58; southern packet

lines for, 9–10, 52–54, 62, 72–73; steamships carrying, 117. *See also* textile industry
"cotton triangle," 54, 59, 82
Cram, J. Sergeant, 141, 147
cranes, container, 182, *183*
credit system, 56–58
crime: organized, 11, 12, 167–68, 171–72, 175–76; waterfront, 124–25, 127, 167
Cronin, William, 10
Crooper, James, 42–43
Cunard, Samuel, 116–17, 119, 120
Cuomo, Andrew, 206
Cuomo, Mario, 196
Customs, US Bureau of, 109, 221n2

Daly, Elizabeth and family, 100
Dattel, Gene, 44, 56
David, Joshua, 200
Day, Dorothy, 171
de Blasio, Bill, 206
Delancey, James, 34
Delancey, Oliver, 34
De Lancey, Peter, 27
Delaware, Lackawanna & Western Railroad, *113*, 114, 159, 163
Delaware and Hudson Canal, 67, 70
Denneut, John M., 58
Department of City Planning, 194, 209–10
Department of Docks, 2, 5–6, 13, 131; budget of, 132; construction by, 102–3, 137–40, 143–51; corruption in, 141–42; employees of, 140–41; land repurchase by, 134–37, 144, 188; master plan of, 133–34; metropolitan district proposal and, 152; revenue of, 140, 146, 155; waterfront controlled by, 76, 106, 131–32; water-lot maps of, 26, 29. *See also* Department of Marine and Aviation
Department of Docks and Ferries, 157, 167
Department of Marine and Aviation, 13, 176; pier construction by, 167, 185, 197; revenue of, 177; revitalization plan of, 180. *See also* Department of Docks
Depau, Francis, 45
derricks, floating, 138, *139*

Dewey, Thomas, 175
Dickens, Charles, 97
dikes, 209
Diller, Barry, 205–6, 210
Dinkins, David, 194, 196
docks. *See* piers; wharves
dockworkers. *See* longshoremen
Doig, Jameson, 155
Dongan Charter, 4, 22, 74
dories, 4
drays, 75, 88, 162
Drennan, Matthew, 178
Driscoll, Alfred, 175
Dunn, Johnny "Cockeye," 169–70
Durst, Douglas, 206
Dutch settlers, 4, 19, 21, 135–36, 207, 222n8

East River, 3, 4, 20–21; ferries on, 29, 30, 74–75; immigrant settlements on, 91, *92*, 93–102, 220n9; piers on, 17, 29–30, 121, 122–23, 128, 129–30, 143, 185; revitalization projects for, 135–36, 180; sewage in, 96; shipping lines up, 8, 10; shipyards along, 30, 47, 85, 117, 193; water-lots on, 5, 25, 26–34; waterway connections to, 63, 64; work and life along, 48–50, 88
Elizabeth (NJ), 156, 177, 179–80, 183–84
Embargo Act of 1808, 40
England: revolution against, 19, 39; steamship companies in, 119–21; trade with, 39–40. *See also* Liverpool (England); London (England)
Erie Basin, 69
Erie Canal, 10, 67, 68–70, 77; barges and ships on, 66, 67, 69, 109; building, 18, 213n18; opening of, 47, 62; passenger lines on, 78
Erie Railroad, *113*, 114, 158–59, 163

Fall River, *64*, 73
ferries, 11, 74–76, 77; Jersey City, 36, 74; leases for, 114–15, 132, 140; Long Island, 29, 30, 74–75
Fisher, James, 174; *On the Irish Waterfront*, 102–3, 140, 172
Five Points, 80, 92, 97
Flack, John, 31, 32
flood protection system, 210–11
flood zones, FEMA, 207, *208*, 209

Ford, Gerald, 187
Ford, Henry, 83
France, trade with, 40. *See also* Le Havre (France)
Freedom Tower, 7, 14
Friends of Hudson River Park, 197
Friends of the High Line, 133, 200
Frost, S. A., 126
Fulton, Robert, 65
fur trade, 19, 135

Gansevoort Market, 35
garment industry, 85–86. *See also* textile industry
gas, manufactured, 32–33, 70
General Theological Seminary, 35, *36*, 108, 147
George, Sir Rupert, 79, 82
George III, King, 27
German immigrants, 90–91, 220n9; decline in, 104, 106; neighborhoods of, 91, *92*, 93–94; transatlantic passage of, 79, 82–83, 90
Gibbons v. Ogden (US Supreme Court ruling), 66
Ginnetty, Joseph, 202
Glick, Deborah, 205
global warming, 14–15, 206–9
Gotham Court, *94*, 97–102
Gouverneur, Nicholas, 31, 32
grain elevators, floating, 66, 69
Grand Central Terminal, *113*, 114, 149, 150
Great Depression, 166–67, 176
Great Lakes, 63; canal link to, 62, 67, 68, 77
Greene, George S., 140, 144, 146
Greenwich Village, *104*; census block groups in, 201, *203*; gentrification of, 201–4; immigrant settlements in, 91, *92*, 102, 103–6, 220n9; land repurchase in, 135, *136*, 144, *145*; pier building in, 144–46, 147, 180; population of, 204; sanitary districts in, 103, *105*
Griesa, Thomas P., 195
grocery stores, 98
Guardian Angel Catholic Church, 108, 169, 172, 173, 175

Haight, John, 49
Hall, Charles, 32

Hall, Henry, *32*
Hammond, David, 200
harbormasters, 127
Hartog, Hendrik, 21–22, 24–25
Hayes, Joseph, 201, 202
Higgins, A. Foster, 142
High Line, 161, 166, 198
High Line Park, 166, 199–200, *201*
Hintz, Andy, 169–70, 173
Hoag, Sydney, 131, 134–35
Hoboken (NJ), 130, 146, 151, 157–58
Holland, Henry, 27
Holland-America Line, 180, 197
Holland Tunnel, 155, *156*, 200
housing: Battery Park City, 192–93; Chelsea, 107–8; East River, Ward 4, 95–96; Gotham Court, 97–102; Greenwich Village, 103–4, 201, 204; Hudson River Park, 198–99; London Terrace, 108; Pier 40 area, 205; Stuyvesant Town, 14, 33–34, 208; tenement, 30, 89, 91, 97, 169. *See also* neighborhoods
Hudson, Henry, 80, 194, 207
Hudson River, 3, 4, 20–21; development along, 102–3; Erie Canal and, 68–70; ferries on, 36, 74; freight crossing, 157–63, 165–66; immigrant settlements on, 91, *92*, 102–8, 220n9; land repurchase along, 135, *136, 145*; revitalization projects for, 180; sewage in, 96; shipping lines up, 64–68; steamboats on, 78; tunnels under, 153, 155, *156*, 164–65, 200; water-lots on, 5, 6, 25–26, 34–37, 190–94, 214n30; waterway connections to, 8, 10, 62, 63, *64*; Westway proposal for, 14, 195–96, 206. *See also* Chelsea; Greenwich Village
Hudson River Park, 196–99, 203; Battery Park and, *193*; High Line and, 200; island park connection to, 205–6; pedestrian walkway of, 137
Hudson River Park Trust, 133, 196–97, 205
Hudson River piers, 6, 17; abandonment of, 189–90, *192*; building, 86, 180, 185; Chelsea, 36–37, 106, 146–51, 176, 224n29; city-owned, 102–3, 128; Greenwich Village, 144–46; length of, 121, 140, 143, 144–51, 152, 224n29; master plan for, *134*; Pier

40, 180, 197–98, 205; railroad use of, 161, 166; study of, 129; water-lot grants for, 34, 35

Hudson River Railroad, 112, 160, 161

Hunt, Robert, 77

hurricanes: Irene, 209, 210; protection from, 210–11; Sandy, 14–15, 33, 207–9, 210

immigration, 87, 89–93, 104, 106, 220n9, 221n37; chain, 91; labor supply from, 177; nativist restrictions on, 93; transatlantic, 17–18, 77–83, 90, 110, 218nn30, 33. *See also* German immigrants; Irish immigrants

Industrial Revolution, 40; English, 9, 39, 44, 58–61, 83, 216n45; New York City and, 63, 83–86; steam power in, 115; wealth created by, 72

infant mortality, 98

International Longshoremen's Association (ILA), 6, 108; corruption in, 11, 12, 171, 172, 173–74, 175; Joe Ryan and, 169–70; metropolitan district and, 152–53; strikes by, 175, 181–82; Waterfront Commission and, 175–76. *See also* longshoremen

Interstate Commerce Commission, US (ICC), 152, 159

interstate highway system, 13, 179

interstate metropolitan district, 152–53

Irish immigrants, 90–91; neighborhoods of, 11–12, 37, 91, 92, 93–108, 201–2, 203; Newport (RI), 72; number of, 104, 106, 220n9, 221n37; and potato famine, 79, 82, 83; stereotypes of, 101; transatlantic passage of, 79–80, 81–83, 90

Irish waterfront, 81, 90; Fisher on, 102–3, 140, 172; gentrification of, 201–4; movie about, 12, 170–73, 175; neighborhoods of, 12, 19, 93–108. *See also* Chelsea; Greenwich Village

iron works, 68; Colwell, 106; Novelty, 85, 86, 118

island park, 205–6, 210

Jacobs, Jane, 201

James II, King, 27

Jansen, William, 74

Jay, Peter, 27

Jefferson, Thomas, 40, 54

Jersey City (NJ): ferries to, 36, 74; railroads in, 157–58, 159; shipping relocation to, 130, 146, 151

Jesuit priests, 12, 108, 171, 172, 174

Johnson, Malcolm, 12, 170, 171–72, 173

Johnson, Walter, 55

Kazan, Elia, 12, 170, 172, 226n25

Kearney, Patrick, 50

Kefauver, Estes, 173

Kennedy, James, 140

Kerriochan, Hugh and family, 202

Kleindeutschland (Little Germany), 91, 92, 94

labor force: Catholic Workers Movement for, 171; container revolution and, 13, 180–81, 199; day-labor system for, 11, 13, 87–89, 93, 168–69; social reform of, 173–77; Tammany Hall and, 139–41, 167. *See also* International Longshoremen's Association (ILA); longshoremen

Lawlor, Owney, 169

Lazarus, Emma, 110

Leary, George, 141

Le Havre (France): packet lines serving, 8, 40, 45–46, 53, 76, 215n16; transatlantic passengers from, 79, 110

Lehigh Valley Railroad, *113*, *114*, 158–59, 163

Levinson, Marc, 178, 181

Lindsay, John, 181

Little Germany, 91, 92, 94

Liverpool (England), 109; cotton shipments to, 53–54, 55, 56–61, 110, 219n33; exports from, 53, 54; maritime infrastructure at, 59–60, 69, 127; packet lines serving, 8, 9–10, 17–18, 40–45, 48, 76, 115, 215n16; public control of waterfront in, 60–61; steamships serving, 116–19; tides at, 20, 59; transatlantic passengers from, 79, 80, 81–82

Livingston, Robert, 65–66

Livingston, Walter, 77

London (England), 109; imports at, 46; maritime infrastructure at, 127; packet lines serving, 8, 40, 45, 76, 215n16

Long Island (NY), 2, 3, 19, 70; climate change and, 209; ferries to, 29, 30, 74–75
Long Island Sound, 10; inland access via, 62, 63, 64; steamers of, 70–74, 78
longshoremen: exploitation of, 167–69; Irish, 90, 93; Liverpool, 60–61; movie portrayal of, 170–73, 175; neighborhoods of, 11–12, 30, 37, 93–108, 201; stereotype of, 88–89; wages of, 50. *See also* International Longshoremen's Association (ILA); labor force
Lorillard, Peter, 25–26
Louisiana Purchase, 54, 55
Lovett, William, 181

Madden, Owen "Owney the Killer," 167
mail service, transatlantic, 42, 45, 116–17, 119
Malden, Karl, 170, 171
Mandel, Henry, 108
Manhattan Gaslight Company, 33
Manhattan Island, 1–2, 3, 19–21; area of, 1, 5, 7, 31, 34, 35–36, 37, 207, *208*; consolidation of, 75–76, 142–43, 218n22; ferries to, 74–75, 77; manufacturing on, 83–86, 177–78; population of, 63, 90–92, 220n9; West Side of, 158–66. *See also* New York, Port of; New York City
Manhattan waterfront: corruption on, 167–69; crowding on, 86, 110–11, 142, 151, 153, 157, *162*; decline of, 2, 5, 12–13, 176–77, 180, 185–88; future of, 209–11; infrastructure of, 1–2, 4–7, 17, 128; movie portrayal of, 12, 170–73, 175; public *vs.* private control of, 2, 5–6, 16, 21–24, 37, 125–26, 128, 130, 131–33, 161, 188, 197–98, 205–6; rebirth of, 2, 14–15, 133–34, 137–40, 180, 188–89; reclamation of, 134–37; redevelopment of, 14, 189–90, 194–96, 207; reform on, 12, 125–27, 173–77. *See also* East River; Hudson River; New York, Port of
Mannahatta Project, 31
maritime infrastructure, 1–2, 4–7, 17, 128; challenges to, 121–25; English, 59–60, 69, 127; Port Authority projects for, 156, 157, 166, 190–94; railroad, 102, 190. *See also* bulkheads; piers; wharves
Marshall, Benjamin, 41–42
Marshall Plan, 167
Marx, Karl, 44
Mathias, Charles, 79
Maury, Matthew Fontaine, 17
Mazet, Robert, 141
McAdoo, William, 153, 154
McCarthy, Joseph, 226n25
McClellan, George B., 133
McCormack, William "Mr. Big," 172, 174
McDonald, Thomas and family, 201, 202
McLean, Malcom, 13, 178
McSorley, John, 140
Meany, George, 173
Meatpacking District, 203, 204
Melville, Herman, 80–81, 99
Milliner, William, 27
Mississippi Territory, 54, 55
Mobile (AL): packet lines to, 9, 52, 53; shipping volume at, 109, 176–77, 217n57
Montgomerie, John, 22, 27
Montgomerie Charter, 4, 22–23, 24–25, 26, 27, 143
Moore, Benjamin, 35, 36, 37
Moore, Clement C., 35, 36, 106, 108, 147
Morris, Robert, 39
Morris Canal, 67, 70
Morrison, John, 117
Moses, Robert, 189
Mulcahey, Mary, 100
Mulligan, Peter, 48–49
Murphy, Charles F., 141
Murphy, Daniel and family, 100–101
Murphy, John, 140

Narrows, 3, 19, 63
National Archives and Records Administration (NARA) database, 79–80, 81
National Labor Relations Board (NLRB), 173
Naughton, Daniel, 141
neighborhoods, 11–12, 89; Chelsea, 37, 102, 106–8; code of silence in, 11, 174–75; container revolution and, 13; East River, 30, 93–96; ethnic

makeup of, 91–93; gentrification of, 199, 201–4; Gotham Court, 97–102; Greenwich Village, 102, 103–6; Irish immigrant, 80, 92, 169. *See also* housing

New Amsterdam, 4, 19, 21

Newark (NJ): airport in, 155; container facilities in, 13, 177, 183–84, 188, 199; investment in, 156, 179–80; organized crime in, 12; waterfront jobs in, 180

New Jersey: container ports in, 12, 13, 177, 188; investment in, 156, 177, 179–80, 183–85; metropolitan regional plan for, 151–56; railroads based in, 113, 114–15, 151–52, 157–66. *See also* Jersey City (NJ); Newark (NJ); Port Authority of New York and New Jersey

New Orleans (LA): cotton trade in, 47, 57, 59, 217n57; imports at, 46; packet lines for, 9, 52, 53, 54, 76; shipping volume at, 109, 176–77

Newport (RI), 1, 18–19, 64; British occupation of, 19; steamboat service to, 71–72, 73; wealth in, 72

newspapers, 12, 41, 45, 170, 171–72

New York, Port of, 8–11, 38; cotton trade at, 54, 55–56, 59, 73, 217n57; imports and exports of, 21, 39, 40, 46–47, 53, 55–56, 68, 150, 157, 166, 176; infrastructure challenges of, 121–25; inland waterways to, 10, 62–63, 64, 77; labor force at, 87–90, 180–81; maritime dominance by, 45, 61, 62, 76–77, 109, 142, 176; metropolitan model for, 152–56; Newark as center of, 156, 183, 188; packet lines serving, 8, 40, 44–47, 52–54; rail lines to, 111–14; reform of, 128–30; regional extent of, 143, 151, 161; shipping volume at, 109–11, 150–51, 176–77, 178, 221n2, 222n4. *See also* longshoremen; Manhattan Island; Manhattan waterfront; Port Authority of New York and New Jersey

New York and Harlem Railroad, 112, 222n7

New York Central & Hudson River Railroad, 112, 113

New York Central Railroad, 10, 112, 113, 160; freight volume of, 151,

162–63; gridlock caused by, 162; High Line of, 161, 166, 198, 199–200; monopoly by, 159–61, 166; rivals of, 142, 151–52, 157, 158; street tracks of, 163, 166, 199

New York Chamber of Commerce, 51, 125, 142, 154

New York City: British occupation of, 19, 39; budget of, 130, 187; charters establishing, 4, 22–23; creation of greater, 75–76, 142–43, 218n22; eminent domain power of, 137, 144, 146; federal loans to, 187; Great Fire of 1776 in, 19; immigration to, 87, 89–93, 104, 106, 220n9, 221n37; industrialization of, 83–86; near-death of, 185–87, 188; Parks Department of, 197; piers owned by, 102–3, 128–30, 142; population of, 1, 21, 186; revitalization plan of, 14, 194–96; rival seaports of, 18–19, 38–40; service economy of, 192–93; Sinking Fund of, 129, 132; street-cleaning laws of, 23–24; waterfront control by, 22–23, 128, 131, 188. *See also* Department of Docks; Department of Marine and Aviation; Manhattan Island; Manhattan waterfront; water-lots

New York Gaslight Company, 31–33

New York Harbor, 1, 16; dredging, 126–27; geography of, 2–4, 3, 18, 19–21, 20, 62; Waterfront Commission of, 175–76, 180–81. *See also* New York, Port of

New York Public Library, 150, 206

New York State: metropolitan regional plan for, 151–56; public authorities in, 189; Saxe Law of, 163; waterfront control by, 125, 128, 131, 143

New York Sun, 12, 170, 171–72

North River Steamboat Company, 65–66

Nott, Eliphalett, 32, 33

Novelty Iron Works, 85, 86, 118

ocean liners, 2; decline of, 13, 180; *Lusitania* and *Titanic*, 147; piers for, 106, 149. *See also* steamships

O'Connell, James and family, 100

O'Donnell, Monsignor John J., 169, 173

O'Dwyer, William, 179
On the Waterfront (movie), 12, 170–73, 175
Orr, Isaac, 126
O'Toole, James and family, 201, 202

packet ships: Black X Line, 45, 116; Blue Swallowtail Line, 8, 45; building, 30, 47, 77; canal, 78; crew of, 48–52; Dramatic Line, 8, 117, 219nn33; Holmes Line, 53; Hudson River, 64–68; New Line, 8; Old Line, 45–46; passengers on, 77–83, 110, 218nn30, 33; piers for, 122; Red Star Line, 8, 45, 219n33, 224n29; Red Swallowtail Line, 45; rival lines of, 44–46, 215n16; schedules for, 40–42; Second Line, 46; southern, 9–10, 52–54, 72–73, 76, 117; transatlantic, 8–9, 17–18, 40–47, 48, 62, 76, 116–17. *See also* Black Ball Line; cotton; sailing ships; steamships
parks, 194–95; Battery, 7, *193*; Battery, 196; Bryant, 206; Central, 206, 222n8; High Line, 166, 199–200, 201; island, 205–6, 210. *See also* Hudson River Park
Pataki, George, 196
Patterson, Victor, 169
Pennsylvania Railroad, *113*, 114, 130, 158, 163
Philadelphia (PA), 1, 18; British occupation of, 19; freight traffic to, 130; imports at, 46; port fees in, 126; shipping volume at, 109, 110, 176–77
Philadelphia & Reading Railroad, 158–59
Piano, Renzo, 200
Pier, Stephen, 77
piers: abandonment of, 13, 124, 177, 180, 185, *186*, 188; building, 2, 4, 5, 16, 21, 23, 26, 38; city-owned, 102–3, 128–30, 142; crowding at, 110–11, 142; East River, *17*, 29–30; leases for, 140, 146, 150, 224n29; length of, 121–23, 134, 143–46; master plan for, 133–34; Port Newark, 179; public control of, 5, 131–32, 134–37; railroad use of, 114–15; rebuilding, 137, 139–40, 143, 167; regulation of, 125–26; wet basins *vs.*, 60. *See also* Hudson River piers
Pig Alley, *104*, 175

Place, Ira, 160
Pollock, Carlyle, 26
Port Authority of New York and New Jersey, 128, 132–33; infrastructure projects of, 156, 157, 166, 190–94; New Jersey investment by, 156, 177, 179–80, 183–85, 188. *See also* New York, Port of
Port of New York Authority, 153–56
Priest, George, 79
Providence (RI), steamboat service to, 71, 74, 78
public authorities, 189; Battery Park City Authority, 191–92, 194; Hudson River Part Trust, 196–97; redevelopment by, 189–90. *See also* Port Authority of New York and New Jersey
Pulling, Ezra E., 96, 98, 99, 100

Queens (NY), 75–76, 142–43, 218n22

Railroad Administration, US, 153
railroads, 111–14; elevated, 161, 166, 198, 199–200; freight volume of, 150–51, 167, 178; industrialization and, 83–84; infrastructure for, 102, 190; leases for, 102, 114–15, 135; New Jersey–based, 114–15, 151–52, 157–66; regulation of, 151–53, 159, 163; steamships *vs.*, 68, 71, 130, 142; West Side problem of, 163–66. *See also* New York Central Railroad
railways, horse-drawn, 75, 163
Rapid Transit Commission, 163. *See also* subway system
Remick, Joshua, 48
Remsen, Henry, 27
Remsen, Joris, 27
Riis, Jacob, 99, 101–2
Riley, John F., 168
Robbins, Tom, 195–96
Rockefeller, David, 191–92
Rockefeller, Nelson, 192
Roosevelt, Franklin, 167
Ruggles, Samuel, 69
Russell Sage Foundation, 87, 88–89, 90, 153–54
Rutgers, Anthony, 27
Ryan, Joseph, 169–70, 172, 173, 175

sailing ships, 16–19, 109–11; clippers, 47, 62, 115; *Empress of China*, 39;

Long Island Sound, 70–71; *May-flower*, 17; passengers on, 18, 110; periaugers, 74; schooners, 70–71, 110; sloops, 64–65, 70–71; steam power and, 85, 115–16. *See also* packet ships
sailors, 48–52; fleecing of, 167; impressment of, 40; wages of, 50–52
Sailor's Magazine, 49, 51
Saint, Eva Marie, 170
Sanderson, Eric, 31
Sandy, Superstorm, 14–15, 33, 207–9, 210
sanitary districts, 95–96, 103, *105*
Savannah (GA): cotton from, 217n57; packet lines to, 9, 52, 53, 76; shipping activity at, 109, 110; steamship to, 115–16
Schermerhorn family, 135–36
Schulberg, Budd, 12, 170, 171, 172, 174
sea level rise, 209
Seamen's Friend Society, 49, 51
sewage. *See* waste removal
Sheen, Bishop Fulton J., 174
ships. *See* barges; container shipping; ocean liners; packet ships; sailing ships; steamships
shipyards, East River, 30, 47, 85, 117, 193
Shkuda, Aaron, 192
Shumer, Chuck, 206
Sierra Club, 195
Silver, William and family, 202
Simonetti, Nick, 181
Slater, Samuel, 72
slavery, 54–55, 57–58
Smith, Alfred, 94–95, 100
Smith, Henry, 50
Smith, Tanner, 167
Smithers, Henry, 60
Sneezy, Carrie and family, 202
South, America: African American migration from, 186; imports of, 53; shipping routes to, 8, 9–10, 52–54, 72–73, 76, 109, 221n2. *See also* cotton
South Street: bulkheads along, 135, 137; piers along, 122, *123*; water-lots to, *28*, 29–30; widening, 137, *138*, 140
Spellman, Cardinal Francis, 174
Staten Island (NY), *2*, *3*, 19; city consolidation with, 75–76, 142–43,

218n22; climate change and, 209; waterfront expansion in, 151
Statue of Liberty, 110
steamships, 2, 10, 115–19; advancements in, 58, 77, 119–21, 147; British, 119–21; building, 85; Cunard Line, 106, 117–18, 120–21, 122, 130, 222n17, 224n29; Fall River Line, 73; ferry service by, 74–75; Great Western Steamship Line, 116; Hudson River, 65–68; inland routes for, 78; Long Island Sound, 70–74; Morning Line, 66; passengers on, 110, 118–19, 150; People's Line, 66, 71; piers for, 102, 106, 121–23, 142, 143–44, 146, 147, 149–50, 224n29; safety regulations for, 118–19; sailing ships *vs.*, 85, 115; shipping activity of, 109–11; speed of, 116, 120, 222n17; Starin Line, 114; transatlantic service by, 116–19; Union Army leasing of, 112; Union Line, 66; United States Steamship Line, 117–19; White Star Line, 106, 150, 224n29. *See also* packet ships
steerage passengers, 79, 219n33
stevedores. *See* longshoremen
St. John's Terminal, *198*, 199, 210
Stone, Moses, 48
storm barriers, 210
streets, 6, 23; construction of, 4, 5, 25–26, 31; paving and cleaning, 23–24; rail tracks on, 163, 166, 199. *See also* South Street; West Street
Stronger, More Resilient New York, A, 210–11
Stuyvesant Cove, 7, 14, 30–34, 208
Stuyvesant family, 31, 222n8
St. Veronica's Catholic Church, 103, *104*
subway system, 163; flooding of, 207; freight, 164–66; PATH, 153; rebuilding, 196
Sullivan, John, 100
Sullivan, Thomas, 140

Tammany Hall, 218n22; labor force and, 139–41, 167; opposition from, 125, 128, 131, 152–53; Ward 4 and, 94, 95
textile industry: English, 9, 39, 44, 55, 56, 59, 83, 85–86, 216n45; French, 45–46; New England, 72–73, 74, 86

Thompson, Francis, Jr., 79
Thompson, Francis, Sr., 41–42, 79
Thompson, Jeremiah, 41–43
Thompson, Samuel, 77
Tompkins Square, 91, 92
Torey, William, 108
trucking, 13, 155, 156, 178–79
tugboats, 158–59

Upper Bay, 3, 20–21

Van Cortlandt, Augustus, 27, 34
Van Cortlandt, Peter, 27
Van Dam, Rip, 27
Vanderbilt, Cornelius, 72, 74; railroads
    of, 111–14, 115; steamboats of, 65–
    66, 71, 77, 112
Ver Planck family, 135–36
Vision 2020, 209–10
von Furstenberg, Diane, 205–6, 210
Vonhasselen, Otto and family, 202
Vredenburgh, John, 27

Waldman, Stuart, 103
Wallace, Mike, 83
Wall Street, 75, 157
Ward, Nathan, 170
Warren, Whitney, 149, 150
waste removal, 23–24; East River area,
    96; Hudson River area, 103–4, 107;
    New York Harbor, 126–27
Waterfront Commission of New York
    Harbor, 175–76, 180–81
water-lots, 2, 4–5, 22–26; Battery Park
    City, 190–94, 207; charter defining,
    4, 22–23, 24–25; Chelsea, 35–37; East
    River, 26–30; Hudson River, 6, 34–
    35, 214n30; private control of, 121–
    26, 128; public reclamation of, 134–
    37, 144–46, 147, 148; Stuyvesant
    Cove, 30–34; title search of, 133
water system, "Croton-water," 107
waterways, 8–11; inland, 10–11, 47,
    62–63, 64; New York dominance of,
    38–40, 76–77; southern, 9–10, 52–

54; transatlantic, 8–9, 16–19, 40–47,
    77, 79–83
Watson, Robert, 58
Webb, Joseph, 136
Webb, Lewis, 136
Webb, William, 47
Westervelt, Jacob, 47
Westley, Agnes and family, 202
West Shore Railroad, 113, 158–59, 163
West Side Highway, 13, 14, 185–86,
    195–96
West Street: bulkheads along, 136,
    137; congestion on, 162; construc-
    tion of, 25–26, 107, 135; rail yards
    along, 161, 164; relocation of, 148;
    water-lots to, 6, 34; widening, 137,
    138, 140, 148
Westway, 14, 195–96, 206
wet basins, 60, 67, 69, 127
wharfage: control of, 25, 26, 30, 132;
    raising, 126, 128; revenue from, 140
wharfingers, 125, 126
wharves: building, 2, 4–5, 16, 21, 23,
    25–26; East River, 29, 31; Hudson
    River, 6, 34, 86; regulation of, 125–
    26; repair of, 5, 129; wet basins vs.,
    60. See also piers
White, C. P., 136
Whitney Museum of American Art,
    200
Whitson, Margret, 50
Wilgus, William J., 152–53, 163–66
Williams, John, 49–50
Williams, Mora, 50
Wilson, Woodrow, 153
Wood, Silas, 97
workforce, waterfront, 89–93. See
    also longshoremen
World Trade Center, 7, 14, 156, 190–
    94
Wright, Isaac, 41–42
Wright, William, 41–42
Wright brothers, 178

Xavier Labor Institute, 171, 174